CISM COURSES AND LECTURES

The series presents lecture notes, monographs, edited works and proceedings in the field of Mechanics, Engineering, Computer Science and Applied Mathematics.
Purpose of the series is to make known in the international scientific and technical community results obtained in some of the activities organized by CISM, the International Centre for Mechanical Sciences.

INTERNATIONAL CENTRE FOR MECHANICAL SCIENCES

COURSES AND LECTURES - No. 366

NEW DESIGN CONCEPTS
FOR HIGH SPEED AIR TRANSPORT

EDITED BY

H. SOBIECZKY
DLR GERMAN AEROSPACE RESEARCH ESTABLISHMENT

 Springer-Verlag Wien GmbH

Le spese di stampa di questo volume sono in parte coperte da contributi del Consiglio
Nazionale delle Ricerche.

This volume contains 139 illustrations

In order to make this volume available as economically and as
rapidly as possible the authors' typescripts have been
reproduced in their original forms. This method unfortunately
has its typographical limitations but it is hoped that they in no
way distract the reader.

ISBN 978-3-211-82815-1   ISBN 978-3-7091-2658-5 (eBook)
DOI 10.1007/978-3-7091-2658-5

PREFACE

At a time when the increase in global traffic suggests a need for innovative solutions, this book offers a collection of contributions to the design of methodologies for a new generation of high speed transport aircraft, and of supersonic craft in particular.

The contributors come from university aerospace departments, the aircraft industry, and an aerospace research institution: the University of Colorado and Pennsylvania State University in the United States of America, Daimler Benz Aerospace and the DLR German Aerospace Research Establishment in Germany. They have been selected to provide a balance between the practical requirements for development and the tools and concepts for achieving design goals.

The book consists of twenty chapters arranged in three parts:

The first six chapters, after exploring the market outlook, present the challenge of developing the technologies needed to create solutions to high speed air transport within the framework of a variety of economic, environmental, and other practical constraints. Chapter 1 discusses the prospects for the development of a supersonic transport, including unconventional solutions. Chapter 2 provides a method for predicting future aircraft pricing. Chapter 3, outlines a multidisciplinary approach to the development of new aircraft, while Chapter 4 presents the problem as a multipoint design challenge. Chapter 5 lists the technologies needed. The final challenge, certification of a new supersonic aircraft, is examined in Chapter 6.

A collection of design tools follows, with theoretical models of the aerodynamics supporting generic aircraft shape definition, for use in systematic optimization strategies. While aerodynamics clearly dominates in the contents of this book, structural and thermal loads are treated as well in Chapters 7 through 15, stressing a careful selection of design parameters based on mathematical modeling, and reviewing recent techniques for optimization. Chapter 7 introduces phenomena-based tool development, illustrating the value of a detailed understanding of flow phenomena in the transonic flight regime. Chapter 8 then provides mathematically defined supersonic configurations. Both phenomena and configuration models serve as the basis for the geometry preprocessor software development in Chapter 9. The inverse aerodynamic problem formulation of Chapter 10 is compared with other strategies for optimization in Chapter 11. A combination of these techniques is reported in Chapter 12. Thermal problems are discussed in Chapter 13, and structural problems in Chapter 14, applying inverse and optimization strategies. Finally, a global approach to multidisciplinary inverse design and optimization in a parallel computing environment is treated in Chapter 15.

In the last five chapters various knowledge bases are used for special and innovative aircraft concepts. Comparison of optimum conventional and novel configurations resulting from systematic design approaches stimulates the designer's creativity, so that he can improve on his own methods. Certain aspects of the initial challenge are encountered in some case studies in Chapter 16. The industrial use of optimization tools is illustrated in Chapter 17. Chapter 18 discusses the possibility of improving aircraft performance by establishing laminar flow on aircraft components. The concluding chapters are devoted to an unconventional configuration, the oblique flying wing: Chapter 19 investigates a case study, with the application of industrial methods, while Chapter 20 discusses other studies of this unusual aircraft and some of its aerodynamic characteristics.

The collaboration between authors R. Seebass and H. Sobieczky was funded by the Alexander von Humboldt Stiftung with a Max Planck grant which helped to make possible the results outlined in Chapters 7, 8, 9 and 20, and substantially supported the editor in his idea of organizing the lecture series "New Design Concepts for High Speed Air Transport" held at CISM in June 1995. The encouragement of W. Schneider of Vienna is also gratefully acknowledged.

Many thanks go to Michael Klein and Stephanie Alberti of the DLR Institute for Fluid Mechanics in Gottingen, who put all the manuscript data into book form, and to the DLR for providing the computer equipment to make this possible.

H. Sobieczky

CONTENTS

CHAPTER 1

THE PROSPECTS FOR COMMERCIAL SUPERSONIC TRANSPORT

A.R. Seebass

University of Colorado, Boulder, CO, USA

1.1 Introduction

This chapter on the prospects for commercial transport at supersonic speeds must begin by deciding what we will call the generic prospective aircraft. Since the first generation aircraft were called Supersonic Transports, or SSTs for short, this practice is continued here. Today, in the United States, NASA's nomenclature is High Speed Civil Transports or HSCTs, while in Europe and Japan it is Supersonic Commercial Transports or SCTs.

The title of this introductory chapter may seem ill-advised. Commercial transport at supersonic speeds has been a reality since 1976. Indeed, it has been a great technical success. The Concorde fleet has flown over 300,000 hours, most of them at supersonic speeds, and it has done so with over 93% reliability. These aircraft will be in service for many years to come [1]. I can go to my local travel agent and buy a ticket to fly from Kennedy International Airport outside of New York City to Heathrow Airport outside of London on British Airways, or to Charles de Gaulle Airport outside of Paris on Air France, and back. The round-trip fare for the summer season, 1996, was $7,574 for London and $6,516 for Paris. The corresponding first-class, business, and full coach fares are $6,752, $4,496, and $2,274 for London, and $5,700, $3,220 and $2,042 for Paris; the discount coach fares are $586 for London and $838 for Paris. The cost of halving my flight time between New York and London or Paris is, averaging the two trips, about 113% that for first-class, 183% that for business class, 425% that for coach and nearly 10 times that for discount coach. During the previous winter season, the discount coach fares were about 50% less, making Concorde travel over 15 times more expensive than discount coach then. The discount fare to London during the 1996-97 winter season was less than 1/30 of the Concorde's $7995 fare there.

We can probably assume that this fare is covering the direct operating cost of the Concorde, exclusive of the depreciation or amortization of the aircraft itself. At these fares the market for supersonic travel is very limited.

Current scheduled Concorde flights include London - New York, Paris - New York in the summer, and London - Barbados (weekly). Recent reports on the Concorde indicate that the dozen now in service are under-utilized [2], [3]. Excursion flights are a small but growing part of the Concorde operations. While service to and from Dulles Airport to de Gaulle and to Heathrow was provided by both airlines for many years, this (from Dulles to Heathrow) was discontinued in November 1994.

The first SST to fly was the Tupolev-144, with its maiden flight on December 31, 1968, a year before the Concorde's first flight. Tu-144 mail service began on December 26, 1975. Passenger service commenced on November 1, 1977, but was discontinued 7 months later. While this aircraft was not an operational success, the Concorde has been an operational success for the two airlines that operate this small fleet. Commercial transport at supersonic speed is a reality. Does a second generation SST make sense? This chapter reviews the Concorde and U.S. SST programs, and provides the author's own conclusion regarding the prospects for a second generation SST. The readers should develop their own conclusions; this book will help them to do so.

1.2 The Concorde

On November 5, 1956, the British had their first meeting of the Supersonic Transport Aircraft Committee, or STAC. The members had concluded that the U.S. Boeing 707 and Douglas DC-8 would capture so much of the subsonic market for commercial aircraft that the only options available to them were to go above the speed of sound or to give up the market [4]. It may have been better strategy to remain with subsonic aircraft, although the Concorde program did much to bring Britain into the European community.

In March 1959 STAC urged the controller of aircraft in the Ministry of Supply to consider the development of a supersonic transport, estimating a market of 125-175 aircraft. The British then approached the French about a joint program, with one goal being their eventual admission to the European Common Market, then dominated by France. Later there were repeated attempts by Britain to cancel the Concorde. Then President de Gaulle stood by the simple, irrevocable, two page treaty between the United Kingdom and the French Republic, entered into on November 29, 1963 [4], [5].

Commercial flight operations began twenty years ago in January, 1976, with British

Airways (then BOAC) flying between London and Bahrain, and Air France operating between Paris and Rio de Janeiro [6]. In a carefully considered (and in retrospect, enormously wise) decision, Secretary of Transportation William T. Coleman, on February 4, 1976, permitted limited scheduled flights of the Concorde into the United States, initially for a trial period of 16 months [5]. Two flights per day for each carrier were to be allowed into Kennedy, and one flight per day for each carrier was to be allowed into Dulles. Because the FAA operated Dulles, there was no difficulty in obtaining permission to operate there, and commercial service began at Dulles on May 24, 1976. The New York Port Authority banned such flights in March 1976, but this ban was overturned in court and commercial operations began there on November 27, 1977.

To my knowledge there have been no lasting complaints of concern about Concorde operations in selected U.S. airports. But one must presume that for an economically successful SST, the fleet size will not be small, and with this increased SST traffic, it may be necessary to adhere to the latest airport noise level regulations for subsonic aircraft. Perhaps some modest deviations for SSTs will be allowed.

Perhaps the golden age of the Concorde was in 1987 and 1988 when over 60,000 passengers were transported by each airline, more than 40,000 of those in a destination market with load factors just over 60%. In January, 1993, *Air and Cosmos/Aviation Magazine* wrote, "Since 1989-1990 the situation has declined to the point of Air France not even reaching 40,000 total passengers last year. And the results for the first trimester of 1993 do not indicate a substantial increase...." [3].

Important national goals were achieved by the Concorde program. Perhaps the most important was the development of a successful European community aircraft consortium. It is unknown, and not knowable, whether the joint British-French venture to develop the Concorde was the best or the only route to this end. It was achieved, however, and this must be attributed, at least in part, to this joint venture. The French also gained a considerable technological advance in their aircraft. Together, they proved the reliability and safety of public transport at supersonic speeds. The program's cost, through March 1976, was put at between 1.5 and 2.1 billion in 1976 pounds sterling, or between 3.6 and 5.1 billion in 1977 U.S. dollars (yearly weighted exchange rates) [7].

1.3 The U.S. SST Program

The U.S. SST program began in June, 1963 when President Kennedy, in a commencement speech at the Air Force Academy, said, "As a testament to our strong faith in the future of air power.... I am announcing today that the United States will commit itself to an important new program in civil aviation ... a plane that will move ahead at a speed faster than Mach 2, more than twice the speed of sound, to all corners of the globe." The day before this speech the president of Pan American World Airlines had made the announcement that Pan Am was taking options on *six Concordes.* Prior to that Air France and British Airways had ordered eight Concordes each.

A few days later President Kennedy followed up his commencement address with a message to Congress in which he said, "In no event will the government investment be permitted to exceed $750 million" [8]. Development costs were then estimated to be approximately $1 billion.

This program soon became one with two competitive aircraft designs, one by Lockheed and the other by Boeing, and two competitive engine designs, one by General Electric and the other by Pratt & Whitney. Boeing and General Electric were the eventual winners of this competition with the Boeing 2707-100, a swing wing, $M = 2.7$, 200-300 passenger aircraft with a presumed range of 3500 nautical miles, weighing 750,000 pounds, an aircraft that was not then - and perhaps is not now, technically realizable. The swing wing provided both airport noise reductions and improved aerodynamic performance at lower speeds. The weight of the mechanism used to pivot the wings resulted in unacceptably low range, or low payload, or both. The Boeing design evolved to a fixed wing, titanium aircraft, not unlike that proposed by Lockheed. The government's investment in the SST program was to be repaid by royalties on aircraft sales. The government's investments, including interest, would be recovered with the delivery of the 300th aircraft.

The two principal issues of concern with SSTs in the late 1960s were their economic viability because of a likely restriction to subsonic operation over populated areas and airport noise levels upon takeoff. There was limited concern before 1970 about the effects of such aircraft on the stratosphere.

The U.S. program died in the Senate in May 1971, in part from concerns about noise in the airport environs, in part from concerns about its impact on the stratosphere, in part due to politics, and in part because its economic success seemed far less than certain. Today, twenty-five years later, these remain legitimate concerns.

1.4 Air Traffic

The current trends in air traffic are well known [9]. Growth has been positive for most of the past twenty-five years. International travel is growing faster than developed countries' domestic travel, leisure travel is growing faster than business travel, and Asia-Pacific traffic has the largest regional growth rate. Air travel has become a commodity in the following sense: 40% of the travel is discount coach travel; the remaining 60% of the travel is comprised of 20% coach, 30% business class and 10% first class. One would be wrong to conclude, however, that full-fare passengers comprise 60% of travel; most of this travel is also discounted. Because of frequent flyer upgrades and business and other traveller discounts, less than 30% of the passengers on international routes pay "full" fare. In 1995, 95% of the revenue passenger miles in the U.S. were sold at a discount. In the first seven months of 1996, the discount averaged 68%.

Two airline systems have now developed. One is the airline system that dominates most markets and provides air service to both the economy and business passengers, subsidizing economy travel by higher fares for the business traveller. The other provides a true commodity service: no advanced seat assignments, no meals, and sometimes no baggage connection to other airlines. The latter airlines have enlarged the market for commodity travel. For any new aircraft to succeed in the commercial aircraft market, it must compete either in convenience / comfort, or in fare, or some combination of the two.

In 1968 nearly eight million international passengers arrived at or departed from Kennedy International Airport, with 97 thousand arrivals and departures. In 1982 over eleven million passengers arrived at or departed from Kennedy. Because of the introduction of widebody aircraft, this travel was accommodated with under 55 thousand arrivals and departures. In 1993 fifteen million international passengers used Kennedy, requiring 92 thousand arrivals and departures. Once again aircraft arrivals and departures there are close to the airport's capacity.

Expected growth in air traffic cannot be accommodated for long with the world's current airports and aircraft. In developed countries there are few airports that can be added. Thus, it is presumed that some of the increased traffic will be accommodated by larger aircraft. One SST configuration, a wing with passengers inside, flying obliquely, must be large and responds to both the SST and the large aircraft market. This Oblique Flying Wing is discussed in two chapters in this book.

1.5 Market

Within a few months of the first flight of the French and the British Concorde prototypes (March 2, and April 9, 1969), the US SST finalist, the Boeing 2707, had booked 122 options from 26 airlines to purchase aircraft; the Concorde had booked 74 options from 16 airlines. Thus, nearly 200 SSTs were "on order." A year later, in 1970, the FAA predicted 500-800 SSTs would be in operation by 1990. It is now 1996.

Twelve Concordes operate today with a limited schedule and at load factors below 50%. These aircraft need only pay their operating costs exclusive of the amortization of their purchase; they were essentially free to the two airlines flying them [10]. What happened? The fares required to pay for their operation deter their use. Maintenance costs are said to be seven times those of a 747 and fuel costs per passenger mile at least three times that of the 747.

Studies by Boeing and by McDonnell Douglas predict a market for 600 to 1500 SSTs [11], [12]. Mizuno of Japan Aircraft Development predicted a market for 600 Mach 2.5 SSTs with a 5500 nautical mile range, and estimated perhaps a 50% increase in this market derived from its stimulation by the travel time saved [13]. Davies, on the other hand, found it to be between 9 and 36 aircraft, depending on how optimistic one is [14]. The enormous differences among these studies stem from what one projects for the fare required to cover the aircraft's total operating costs. It takes a long time to sell one thousand aircraft. The first Boeing 747 began commercial flights in 1970; twenty-four years later one thousand 747s had been delivered.

The challenge is to design, build, certify and operate an SST while providing the airlines a return on investment comparable to a similar investment in subsonic aircraft. This can only be accomplished with marginally increased fares over those for subsonic transport. The marginal increase in fares required, however, depends upon many factors, including aircraft price and operating cost.

Marginally increased fares - what does that mean? Assume such transport effectively saves the traveler some fraction of a day, or at most, a whole day. Whatever that traveler's expenses would be for that day, or, correspondingly, whatever his income might be for that day, provides a reliable guide as to what he would be willing to pay to save a fraction of a day of business travel, or have as extra time for his vacation. This intuitive judgment agrees with studies which predict little fall-off in ticket sales for a 10% surcharge [11], [13].

As noted earlier, non-discount passengers comprises 30% of the international market. To secure a significant fraction of this market an SST will need to provide three-class service. Current Boeing studies reflect this, but show an SST with about 9% of the passengers in first class, 19% in business class, and 72% in economy. Can an SST succeed if it fills empty seats with discount coach passengers? Can it succeed if it does not?

A final comment is warranted on the growth of revenue passenger miles accorded air transports. The "information highway" will reduce business travel needs. For a few hundred dollars you can buy the software needed for your group to discuss and share visual information by electronic mail. It is now possible, with more expensive software, to have the real-time image of each member in a working group displayed, hear their voices, and share visual information. A telecommunications vice president recently told me that he spent $23,000 on hardware and software and saved $100,000 in travel costs in the first year. The importance of this change was noted some years ago by Simpson in his remarks to the 1989 European Symposium on Future Supersonic-Hypersonic Transportation [15]. When the information highway becomes an international highway, which it now nearly is, this will reduce the need for international business travel while simultaneously expanding the amount of international business. It seems likely that these two effects will offset one another.

Technology has progressed steadily since the Concorde was conceived. But reduced energy efficiency, the sonic bang, engine emissions, and airport noise, remain deterrents to the economic success and acceptability of an SST. Let me now turn to the environmental barriers facing a future SST.

1.6 Environmental Barriers

As I've noted earlier, the U.S. SST program was canceled in part because of environmental concerns. The Concorde's economics have been greatly affected by being prohibited from supersonic flight over most land areas, and by the cost of fuel. The environmental, and thereby political, barriers to a successful SST are: energy consumption, sonic bang, atmospheric impact, and airport noise.

1.6.1 Energy Consumption

The fuel consumed by SSTs per passenger mile is several times that of subsonic transports. Supersonic flight entails a new penalty, that of wave drag. Lift has to equal, and sometimes exceed, weight if there is to be air travel. Wave drag due to lift is inescapable except for an infinitely long swept wing, best approximated by the way, by an oblique wing. Volume can be moved through the air supersonically with no wave drag, but at considerable expense in skin friction drag from extra surfaces.

Sixty countries have ratified a treaty that commits them to better manage their generation of greenhouse gases [16]. Developed countries are to provide plans by the end of this century that show how they will return to 1990 levels of greenhouse gas generation. Does this argue against an SST? As Secretary of Transportation Coleman said in his decision to let the Concorde operate: "It would border on hypocrisy to choose the Concorde as the place to set an example ... (for energy efficiency) while ignoring the inefficiency of private jets, cabin cruisers, or an assortment of energy profligates of American manufacture" [5].

The Concorde achieves 17 seat miles per gallon and, at 67% load factor, is equivalent to a car with only the driver, achieving 12 miles per gallon. But the Concorde's passengers are going more than twenty times as fast and following nearly a straight line to their destination. A future SST should not be rejected because of energy considerations. However, its economics and thereby its market, are more sensitive to fuel costs than its subsonic counterparts and these are not only variable, but jet fuels may eventually be taxed for their carbon content.

1.6.2 Sonic Bang

Just as wave drag due to lift is inescapable, so is the sonic bang. Adolf Busemann liked to illustrate this by depicting the conical shock wave system and its reflection from the ground as the crow-bar that supported the weight of the aircraft [17]. Ironically, while the weight of the aircraft is to be found in the integral of the pressure signature over the ground, it is not to be found in the first-order pressure field there [18]. In the U.S. we call the sonic "bang" the sonic "boom." The "bang" in the sonic boom derives from the abrupt pressure increases through the two, and sometimes more, shock waves emanating from a supersonic aircraft. We call the integral of the positive phase of the pressure with respect to time the "impulse". The bang is directly related to the outdoor annoyance of animals and humans; the impulse is related to structural damage and, to some degree, to indoor annoyance.

The increasing acoustic impedance (i.e., the product of the density and the sound speed) below the aircraft in a real atmosphere freezes the shape of the pressure signature before it reaches the ground. In the approximation of an isothermal atmosphere this occurs in $\pi/2$ atmospheric scale heights, or about 40,000 feet. This knowledge set me and my colleague Al George to tackle the minimization of various parameters of the sonic boom signature, including its bang and its boom, or any weighted average you might use of the parameters. Indeed, for the cruise characteristics of the Mach 2.7 Boeing 2707 at 60,000 feet lifting 600,000 pounds, an aircraft 527 feet long need not have a sonic bang at all, i.e., the pressure field below the aircraft need not steepen into shock waves [19]. But as we noted then, reducing or eliminating the "bang" in the sonic boom increases the impulse, or total pressure loading, for obvious reasons: the bang part of the boom, that is the shock waves, dissipates the energy in the signature. Consequently, reducing or eliminating the shock waves makes the impulse worse.

Very considerable studies by the NASA over the past decade have explored whether or not such shaping of the sonic boom signature would lead to an acceptable sonic boom. The NASA's conclusion reinforces ours of two decades ago. Unless a supersonic aircraft is very light, but long, its sonic boom cannot be reshaped to be acceptable [20]. Very small supersonic aircraft, such as a corporate supersonic transport, may have an acceptable, indeed nearly inaudible, sonic boom. This stems, in part, from a thickening of the shock waves as their strength is reduced.

SSTs will be constrained to subsonic operation over populated areas, and perhaps to supersonic operation over the oceans alone. The penetration of the pressure field of sonic booms into water, versus their reflection from it, is now well understood [21]. For aircraft traveling less than the speed of sound in sea water, this is simply a travelling source of acoustic radiation. Commercial transport at supersonic speeds over the oceans, and perhaps over unpopulated areas, is likely to continue to be acceptable. Flights over land areas with significant populations of wildlife may not be allowed. Through constraints on aircraft routes we can avoid the problems caused by sonic booms, but in doing so we reduce the market for a second generation SST.

1.6.3 Atmospheric Impact

Whenever we burn hydrocarbon fuels using air, we impact the atmosphere and, in some cases, the local air quality. Whatever fuel we burn using air will produce oxides of nitrogen. A concern during the late 1960s was the effect of water vapor from SST engine exhausts on stratospheric ozone levels. It was soon realized, however, that the oxides of nitrogen were much more important [22]. This led the Department of Transportation, in 1972, to launch the Climatic Impact Assessment Program. This monumental and highly regarded 7200 page study, comprising the work of over 500 individuals, concluded that a limited fleet of supersonic transports, such as the 30 Concordes and TU-144s then envisioned, posed an insignificant threat to the atmosphere. This study also aided the extraordinary discovery of the reduction of atmospheric ozone by CFC refrigerants (Freon 11 and 12), culminating in the Montreal Protocol (1987) which will lead to the eventual elimination of these refrigerants.

The oxides of nitrogen catalytically destroy ozone above about 13 kilometers in mid-latitudes; they catalytically create ozone below this altitude. Aircraft emissions are the major unnatural source of these oxides in the stratosphere. They are also an important source of them in the upper troposphere, at least of mid-latitudes in the northern hemisphere [23]. Thus it appears that SSTs in the stratosphere may reduce our protection from ultra-violet radiation by ozone on the one hand. At altitudes of 12-14 kilometers (13 kilometers = 42,650 feet), the effect of these oxides on ozone is minor. The calculated ozone column change due to the injection at 20 kilometers of the amount of NO_x expected from a full fleet of SSTs was about -12% in 1975. New knowledge changed this to +3% in 1979. Since that time, increasing knowledge provided a result of -10% in 1988, about double the -5% predicted ozone depletion if CFC releases remained at their 1974 rate [24]. Recent results show NO_x to be less significant than was once thought, but raise the issue of the effects of engine emissions on stratospheric aerosol surface area. This could also play a role in depleting stratospheric ozone [25].

1.6.4 Airport Noise

Remarkable advances have been made in propulsion since jet engines were introduced. Over the past 25 years there has been about a 20% reduction in the amount of fuel required to produce a unit of thrust [26]. Because much of this gain has come from higher bypass ratios, take-off noise levels have fallen in some cases below those required by current noise regulations. Current SST engine concepts, without augmented suppression systems, are probably 15-20 decibels (equivalent perceived noise decibels) above these standards. Further noise suppression adds weight and reduces thrust. Low lift-to-drag ratios at takeoff demand considerable thrust, and this, in turn, leads to larger exhaust velocities and more noise. At the moment there are sound ideas, but no tried techniques, on how to accomplish this noise reduction with acceptable weight increases. Unlike the sonic boom, however, we are not up against a fundamental momentum balance. A breakthrough is possible. Given that subsonic transport noise levels continue to fall, and the near certainty that conventional supersonic transports will operate only from selected coastal cities, current noise regulations need to be examined to see what airport noise levels might be acceptable from a small fleet of supersonic aircraft.

1.7 The Prospects

The development of a supersonic transport that can be operated at a profit by the airlines, and sold in sufficient numbers for the airframe and engine manufactures to eventually realize a profit as well, remains a challenge. The U.S. and European supersonic research programs now have very focused, and somewhat different, goals. These programs involve the companies that profit from the sale of their subsonic jets. It would take some bold competitive vision, not unlike that which led to the Concorde, for a supersonic transport production program to emerge from these studies. Such an aircraft faces the real possibility that it, too, will be a technical success, but not an economic one. This book, therefore, focuses much of its attention on the underlying tools for the study of such aircraft, as well as on unconventional configurations.

For unconventional configurations the technical and risk barriers are very high. It appears that an oblique flying wing (see Chapters 19 and 20) could provide a Mach 1.4, or higher, transport that operates with a minimum surcharge over future subsonic transports and that competes with them over land as well. If it is large enough it becomes the "New Large Air-craft" and, in this size, such an aircraft may compete in fare with its subsonic counterparts. But without further research, considerable experimentation, and flight tests, this remains a conjec-ture. Such an aircraft would also require rethinking of selected aviation regulations and perhaps even some minor reconfiguration of airports. Both were required with the introduction of the Boeing 747.

A conventional configuration, operating at a higher Mach number, benefits from high productivity and substantially reduced travel times. Because of past and current government research programs, including that which led to the Concorde, the needed research is largely done and the technology mature. Consequently, the development costs of such an aircraft appear to be reasonable. Because of its limited subsonic and transonic performance, and its restriction to intercontinental routes, this aircraft's market is relatively small. As a fleet, its con-tribution to the acoustic environment in and around selected airports may be small enough to deserve continued regulatory relief.

A small, corporate, supersonic transport appears to have a significant market and, if small enough, might well be certified for supersonic operation over land. Military technology and excess production capacity provide the basis for making such an aircraft affordable.

At a meeting on sonic boom research in 1967, Adolf Busemann, having comprehended the concept of bangless sonic booms, concluded this meant we would have to fly in the tropo-sphere to make the sonic boom acceptable. He stood up, placed his arm over his eyes, and said: "This is terrible; we will have to fly through the wind, the sleet, the rain, and the snow." Further research showed even this would not be enough. Large transports will not be able to fly at super-sonic speeds over populated areas.

It may be a long time before most of us can fly twice current speeds at affordable fares. And we may have to fly obliquely to do so. Before this happens, some will have travelled at Concorde speeds in corporate supersonic transports such as the proposed Sukhoi S-21.

1.8 References

[1]   *Aviation Week*
       "Concorde Set to Fly into Next Century," February 12, 1996, p. 39.

[2]   **Quintanilla, C.**

"Unsold Seats Sully Concorde's Snooty Image," *Wall Street Journal*, February 23, 1996, p. B1.

[3]   **De Galard, J.**
Concorde: Le Vrai Bilan Aires 17 Ans, *Air & Cosmos/Aviation Magazine*, No. 1427, May 24-30, 1993, pp. 12-17.

[4]   **Costello, J., and Hughes, T.**
The Concorde Conspiracy, New York: Charles Scribner's Sons, 1976, 302 pp.

[5]   **Horwitch, M.**
Clipped Wings, Cambridge Massachusetts: The Net Press, 1982, 472 pp.

[6]   **Orlebar, C.**
The Concorde Story, London: Temple Press, 1986, 144 pp.

[7]   **Henderson, P. D.**
Two British Errors: Their Probable Size and Some Possible Lessons, *Oxford Economic Papers*, pp. 160-205, 1977.

[8]   **Dwiggins, D.**
The SST: Here It Comes Ready or Not, Doubleday: New York, 1968, 249 pp.

[9]   **Australian Government Publishing Service**
International Aviation, Trends and Issues, Report 86, Bureau of Transport and Communications Economics, 1994, 436 pp.

[10]   **Grey, J.**
The New Orient Express, *Discover*, January 1986, pp. 73-81.

[11]   **Douglas Aircraft Company**
Study of High-Speed Civil Transports, NASA CR 4236, 1990.

[12]   **Boeing Commercial Airplanes**
High-Speed Civil Transport Study, NASA CR 4233, 1989.

[13]   **Mizuno, H.**
Operations and Market, *High Speed Commercial Flight*, H. Loomis ed., Columbus: Battelle Press, 1989, pp. 83-97.

[14]   **Davies, R. E. G.**
The Supersonic Unmarket, *Airways*, September/October 1995, pp. 41-46.

[15]   **Simpson, R. N.**
Assessing the Market for High Speed Travel in the 21st Century, *Du Symposium European sur L'Avenir du Transport Aerien a Haute Vitesse*, November, 1989, pp. 113-121.

[16]   *The Economist,*
Turning up the heat, March 19, 1994, p. 15.

[17]   **Busemann, A.**
The Relationship between Minimizing Drag and Noise at Supersonic Speeds, *High Speed Aeronautics*, Brooklyn: Polytechnic Institute of Brooklyn, 1955.

[18]   **Seebass, A. R. and McLean, F. E.**
       Far-Field Sonic Boom Waveforms, *AAIA J.*, Vol. 6, No. 6, 1968, pp. 1153-1155.

[19]   **Seebass, A. R. and George, A. R.**
       Sonic Boom Minimization, *J. Acoustical Soc.*, Vol. 51, No. 2, 1972, pp. 686-694.

[20]   **Seebass, A. R. and George, A. R.**
       Design and Operation of Aircraft to Minimize Their Sonic Boom, *J. Aircraft*, Vol. 11, No. 9, September 1974, pp. 507-517.

[21]   **Cheng, H. K., and Lee, C. J.**
       Sonic Boom Propagation and Its Submarine Impact: A Study of Theoretical and Computational Issues, AIAA Paper No. 96-0755, 1996.

[22]   **Johnston, H.**
       Reduction of Stratospheric Ozone by Nitrogen Oxide Catalysis from Supersonic Transport Exhaust, *Science*, 6 August 1971, pp. 517-522.

[23]   **Enhalt, D. H., Rohrer, F., and Wahner, A.**
       Sources and Distribution of NO_x in the Upper Troposphere at Northern Mid-Latitudes, *J. Geophysical Res.*, Vol. 97, No. D4, March 20, 1992, pp. 3725-3737.

[24]   **Johnston, H. S., Prather, M. J., and Watson, R. T.**
       The Atmospheric Effects of Stratospheric Aircraft: A Topical Review, NASA Reference Publication, 1250, January 1991.

[25]   **Fahey, D. W., et al.**
       Emission Measurements of the Concorde Supersonic Aircraft in the Lower Stratosphere, *Science*, Vol. 270, October 6, 1995, pp. 70-74.

[26]   **Koff, B. L.**
       Spanning the Globe with Jet Propulsion, William Littlewood Memorial Lecture, May 1991, AIAA Paper No. 2987.

CHAPTER 2

AIRCRAFT ECONOMY FOR DESIGN TRADEOFFS

A. Van der Velden

Synaps Inc., Atlanta, GA, USA

2.1 Abstract

Before the go-ahead is given on the further development of a new transport aircraft design a number of questions need to be answered. The airframe manufacturer needs to know whether it can breakeven on its initial investment easily. The airline will only order this new product if it can expand its market, reduce its cost and increase its revenues. The traveller wants a low ticket price and high comfort. The society as a whole wants this new technology to improve the economy while safeguarding the environment.

Since all of these views are in conflict we can only evaluate a new design by comparing it with existing transports, keeping in mind that the life of a new design can span a quarter of a century or more. The content of this section follows is loosely based on my report [38] "An Economic Model for Evaluating High-Speed Aircraft Designs" of 1989. This report has been updated and reevaluated after my experiences at Airbus.

The present model has been developed to give realistic results to tradeoff engineering features of a design to improve aircraft economy. It is not intended to present an accurate picture of pricing policies of airframe manufacturers and airlines.

The first sections of the paper deal with a market in equilibrium. The economic viewpoints of the manufacturer, airline and passenger are based on the commodity product jet travel is today. These equilibrium market conditions can also be used to make design trade offs for a supersonic transport but we would have to be very careful to infer more. The question of whether such a supersonic aircraft can be sold and at at which price is very much a different issue. Daimler Benz Aerospace and other companies use very refined forecasting models to assess the marketability of a new aircraft and such models are highly proprietary. In the last section of the paper I will present a highly simplified qualitative overview of such a non equilibrium market model

2.2 List of Principal Symbols

C_e	engine price
C_o	operating cost
C_{AFM}	airframe block hourly material cost
C_{ACM}	airframe flight cycle material cost
$C_{L,max,to}$	Maximum takeoff lift coefficient
D_{ac}	aircraft and parts deflator relative to 1994
D_e	engine and parts deflator relative to 1994
D_f	fuel deflator relative to 1994
D_l	labor deflator relative to 1994
H_{EL}	engine labour hours per flight hour
H_{AFC}	airframe flight cycle labour cost (h)
H_{AFL}	airframe block hourly labour cost (h)
i	insurance rate
l	loadfactor
m_{to}	maximum takeoff mass (kg)
m_{af}	airframe mass (kg)
$m_{f,b}$	block fuel (kg)
m_e	engine mass (kg)
n_e	number of engines
P	depreciation period to 10 % of value
R	range (km)
s	number of seats

S	wing planform reference area
T_{max}	maximum uninstalled sea level static thrust
t	time (h)
t_b	block time (h)
$T_{t4,max}$	maximum turbine entry temperature
U	yearly utilization
V	aircraft speed
V_b	block speed (km / h)
W	aircraft weight

Greek Letters

Δt_r	mean time between repairs

Subscripts

ac	aircraft
af	airframe
b	block
e	engine
f	fuel
hs	high subsonic
l	labour
to	takeoff

2.3 Constant Dollar Accounting

The economic model presented in this paper uses four deflators to include the effect of inflation on cost:

1. *The SIC 3721 deflator D_{ac} for the airframes industry.*

2. The SIC 3724 deflator D_e is used for engines and engine parts.

3. The Consumer Price index D_l will be used to deflate labor costs. This is by no means an ac-cuate deflator for all labor involved. Maintenance workers, pilots and the "average" Ameri-can all have different deflators, but the CPI is a reasonable deflator for all these professions.

4. The fuel price deflator D_f

The definition of a deflator is:

$$D_{year = X} = \frac{\text{Price for year=X}}{\text{Price in 1994}} \qquad (1)$$

These deflators are often revised and reliable data is often five years old, they can be found in "Aerospace Facts and Figures" [27]. **Note: All deflators are defined as 1 for 1994.**

2.3.1 Aircraft Cost - Manufacturer

The production cost of for each aircraft in a series will greatly depend on the following consid-erations:

- Aircraft Size. Assuming that all aircraft are designed with a comparable degree of sophisti-cation, irrespective of size, then it is clear that the aircraft cost of development increases lin-early with size. Even though increased research can decrease empty weight, if all designers have a common set of weight to economy tradeoffs they will all stop developing at a compa-rable degree of sophistication. Figure 1 shows relative inflation corrected airframe prices as a function of size. Except for very small aircraft a constant cost per pound can be assumed.

- Technology risk. As Mach number increases, more expensive materials must be used and ex-pensive systems make up a larger fraction of the empty weight which increases the airframe cost per unit weight. Figure 2 shows the relative cost per kilogram airframe as a function of maximum flight speed corrected for inflation. Thought the primary source of this data is mil-itary it can still be used for civil transports because the cost of avioncs and weapons systems was not included. In addition, the Concorde data point was predicted well, as well as the av-erage for subsonic commercial jets.

- Production Size. Based on market research the company has to decide on where we fix its breakeven point. Reference [27] provides sales figures for some successful aircraft programs. For big and medium sized passenger transports a breakeven number of 400 seems reasonable. To obtain the sale price per airframe we now use Wright's 80% learning curve as described in reference [39] and [32] and a representative cost per unit weight and thrust for commercial aircraft. Figure 4 shows actual recorded learning curves.

Based on current pricing policies the following equation can be used to predict the cur-rent (and) future price of aircraft.

$$C_{af} = 7 \left(800 \, D_{ac} \, m_{af} F_0 F_1 \right) n_{be}^{-0.322} \tag{2}$$

$$C_e = 30 \, D_e \, T_{max} \, F_2 \tag{3}$$

$$C_{ac} = C_{af} + n_e \, C_e \tag{4}$$

Where F_2 is shown in Figure 3.

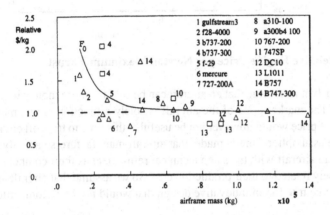

Figure 1   Relative Airframe Price per Kilogram as a Function of Size

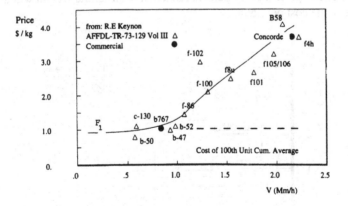

Figure 2   Airframe Price per kilogram as a Function of Speed

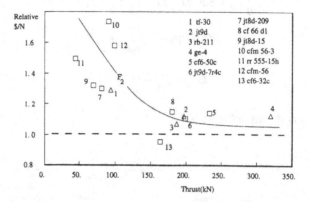

Figure 3   Relative Engine Price per Newton Maximum Thrust

Obviously it is not true that an aircraft can be sold for more money if it is heavier or has bigger engines. The market price of the aircraft is solely a function range, payload and speed. Such definition of price would however not be useful at this time to tradeoff engineering features of a design. The assumption here is made, that aircraft manufacturers can only stay in business if they produce an aircraft with the same technical refinement as their competitors and charge a price that is closely related to their production cost. So we assume that aircraft are a commodity, which would not be true for an innovative design that would not have competitors.

This simple relation should be corrected when non standard materials or technologies are used. Usually the price per pound goes up more than proportional with the introduction of weight saving techniques. This relation will have to be modeled if such weight saving techniques are considered. For composites every percent reduction in structural weight will at least cause a percent increase in the cost per pound.

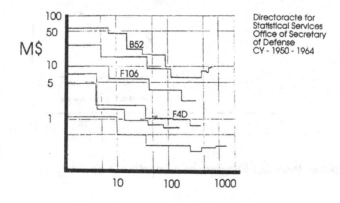

Figure 4   The production Learning Curve for Aircraft

2.3.2 Operational Cost - Airline

Before the airline makes a cost economic comparison it should first evaluate how the aircraft fits in the current fleet in terms of maintainability, passenger comfort, interior arrangement, handling and turnaround time. Reference [37] states that environmental factors such as domestic air policies, expected traffic demand, capacity and frequency, financing and domestic traffic infrastructure also play a role .

It is outside the scope of this section to go into all of these factors but a few are directly related to the aircraft design and should therefore be mentioned:

- The cruise speed is an important criterion since it influences the number of flights per day on a given route. Airlines greatly favor even numbers of flights a day since this enables them to perform the aircraft's (nightly) maintenance in its home port each day. Mach 2 is therefore a good cruise number for the transatlantic range.

- The operational flexibility. Performance in terms of turnaround time, maintenance and take-off field performance should be as good as the aircraft the new aircraft is compared with, to be able to use this method.

1967 the Air Transport Association of America [31] published a standard method to estimate the direct operating cost of aircraft. This method is no longer used to obtain actual direct operating cost, but it is still used to make comparative and parametric studies. Since 1967, the airframe manufacturers have been updating this method to reflect the technological progress. These studies were generally not available to the public. However, in 1978 American Airlines and NASA put out a new study [28] and [33] that reflected the added experience in aircraft operation since 1967.

One of the suggestions in the new method, the introduction of "aircraft related operating expenses" instead of the DOC, was not followed, as far as the author can tell. Since some of the items in the new definition have a large variance for different operations in different nations (e.g. fuel servicing fees, training costs), the author has decided to keep the original definition for the direct operating cost, but to use updated relations from the American Airlines study for the actual equations. The factors in these equations were scaled to represent 1994 European conditions.

To enable us to compare configurations all costs and profits are expressed per seat km.

$$PROFIT = REV - (IOC + DOC) \tag{5}$$

The indirect operating cost IOC will primarily depend on the operator's type of organization and policy. This cost includes maintenance of buildings, servicing of flight operations and administration and sales. In 1985 the average IOC of the world scheduled airlines was 3.5 $ct/pass.km. The direct operating cost DOC include the cost of flying, airplane maintenance and depreciation. Ticket revenue REV depends on the load factor (averaging around 65% of the maximum capacity) and the pricing policies of the airline. In 1985 the average revenue per passenger was 5.7$ct/km, today (1995) it is closer to 8$ct/km

In the following statistical formulae we have used data published in the US Department of Transportation in reference [36].

Defining the block-speed:

$$V_b = \frac{R}{t_{loss} + t_{flight}} \tag{6}$$

The flight time is calculated during the mission integration. The loss time is taken to be 30 minutes. Figure 5 relates the cost of a flight crew, including training and employee benefits, per block hour can be expressed to the aircraft's maximum takeoff weight.

$$C_{o,1} = \frac{750 \cdot D_l \cdot F_3}{V_b} \tag{7}$$

There is an estimated additional 80 dollars for international operations and 190 dollars for supersonic flight per flight hour. The current cost of fuel can be used to calculate the fuel cost:

$$C_{o,2} = \frac{D_f(0.21 \cdot m_{f,b} + 0.74 \cdot t_b \cdot n_e)}{R} \tag{8}$$

As mentioned in the previous section the block fuel includes non-revenue flying and maneuvers.Figure 6 shows that there exists an empirical relation between the the annual utilization of 618 aircraft and the average route block time. In the literature sometimes higher values are quoted, but in practice values over 5000 h are seldom achieved on average. A year has 8760 hours so that means the aircraft would have to be in the air 60 % of its life (including all inspections).

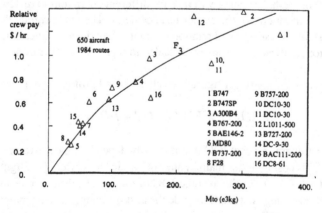

Figure 5   Blockhourly Flight Crew Cost

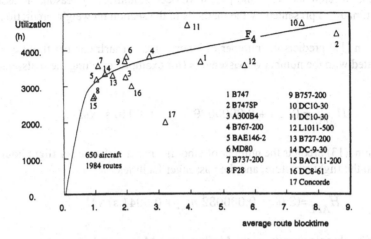

Figure 6   Yearly Aircraft Utilization as a Function of Blocktime

$$U = F_4\,(t_b) \tag{9}$$

We can now calculate the cost of insurance based on an insurance rate i between 0.5 and 1.0%.

$$C_{o,3} = \frac{(C_{af} + n_e \cdot C_e) \cdot i}{V_b \cdot U} \tag{10}$$

The capital cost can be calculated assuming a depreciation to 10% of the original value in period P of 14 years. Four percent of the airframe value is needed for spares, while 30% of the engine value is needed for spares[1]:

$$C_{o,4} = \frac{0.9 \cdot (1.04 \cdot C_{af} + 1.3 \cdot n_e \cdot C_e)}{P \cdot V_b \cdot U} \tag{11}$$

In the published data of reference [36] a large discrepancy can be observed between the above depreciation and the value found in our estimate. It must be noted that in reference [36] the depreciation is made on the original purchase price and not the present purchase price. We can synchronize both methods by correcting the depreciation of the airframe each year with the airframe deflator.

Based on the original ATA'67 publication and the data published by the U.S Department of Transportation, a new model for estimating maintenance cost was proposed. The model is correct for aircraft **five years after their introduction and four years after purchase.** The

---

1.   The total number of depreciation hours cannot exceed the airframe structural life.

original formulae in references [28] and [33] have been simplified by reasonable assumptions and variable estimates as presented by the methods in the section on weight prediction[1].

Equation (13) predicts the number of labor hours per airframe per flight cycle. These cost are associated with the number of passengers (for example: cleaning the seats) and the size of the aircraft.

$$H_{AFC} = (2.14 + 0.0000079\, m_{to} + 0.0046\, s)\, \sqrt{M} \qquad (12)$$

Equation (13) predicts the number of labor hours per airframe per flight hour. This includes repairs to the flight structure, and the passenger facilities:

$$H_{AFL} = (3.08 + 0.000032\, m_{af} + 0.0041\, s)\, \sqrt{M} \qquad (13)$$

For high subsonic operations the Mach number M is set to 1. According to Air France[2] "twice as much effort" is spent servicing the Concorde for every hour of flight as it does for planes that cannot fly at the speed of sound. However, this is primarily due the small size of the Concorde fleet and the square root relations as proposed by ATA will probably be more accurate for larger supersonic fleets. Airframe labor cost are expressed in equation (14) as a function of the labor hours per flight hour and the labor hours per airframe.

$$C_{o,5} = 15 \cdot D_l \cdot \left[\frac{H_{AFC}}{R} + \frac{H_{AFL}}{V_b} \right] \qquad (14)$$

The engine labor cost are related to the mean time between repairs for the powerplant. The mean time between repairs is related to length of the flight cycle (number of takeoffs) and the maximum turbine entry temperature expressed in degrees Kelvin.

$$\Delta t_r = 3604 \cdot t_b^{0.28} \cdot e^{-0.000324 \cdot T_{t4,max}} \qquad (15)$$

The number of labor hours per engine per flight hour can now be expressed as:

$$H_{EL} = \frac{1452 + 0.530 \cdot m_e}{\Delta t_r} + 0.143 \qquad (16)$$

We can now express the engine labor cost in equation (17)

---

1. Instead of using the actual mass of the hydraulics as used in the American Airlines method, we use Torenbeek's expression which relates it to the empty weight of the configuration

2. Peter Thomsen "Concorde Reaches Middle Age" World Press Review June 1990

$$C_{0,6} = \frac{15 \cdot n_e \cdot H_{EL}}{V_b} \tag{17}$$

Airframe parts costs is again divided into the costs per flight cycle and the costs per flight hour. The materials costs per airframe per flight cycle is:

$$C_{ACM} = (20 + 0.00027 \cdot m_{to} + 0.055 \cdot s) \cdot \left[\frac{C_{af}}{C_{af,hs}} \right] \tag{18}$$

$$C_{AFM} = (18 + 0.000077 \cdot m_{to} + 0.069 \cdot s) \cdot \left[\frac{C_{af}}{C_{af,hs}} \right] \tag{19}$$

The materials costs per airframe per flight hour:

$$C_{0,7} = D_{ac} \cdot \left[\frac{C_{AFM}}{V_b} + \frac{C_{ACM}}{R} \right] \tag{20}$$

The engine parts costs is expressed in equation (21):

$$C_{0,8} = \frac{n_e \cdot \left[2.2 \cdot D_e + \frac{0.045 \cdot C_e}{\Delta t_r} \right]}{V_b} \tag{21}$$

In equation (22) the maintenance burden, indirect maintenance costs such as supervision, inventory management are related to the total labor cost:

$$C_{0,9} = 3 \cdot [C_{0,5} + C_{0,6}] \tag{22}$$

Since 1967 the overhead burden has gone up twice as fast as labour inflation. As is clear in Figure 7, a large variation in maintenance costs exists. New aircraft like the B757 and the B767 have much lower maintenance costs than predicted, while some more unusual and older aircraft like the BAE111 have higher maintenance cost. On average the model will arrive at a 24% than actual maintenance costs because so many new aircraft have much lower costs than aircraft that are 5 years or older.

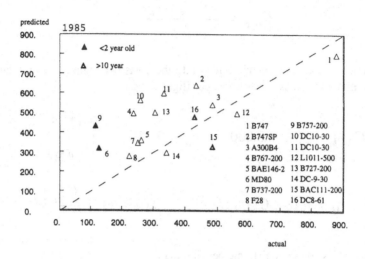

Figure 7   Predicted and Actual Aircraft Maintenance Costs

An estimate of the direct operating cost per seat kilometer can now be obtained:

$$DOC = \sum_{i=1}^{9} \frac{C_{o,i}}{s}$$

(23)

Table 1 shows some published direct operating cost and the costs predicted by this method. A very good correlation is achieved overall.

type	year	seats	R(km)	DOC	DOC-estim.
F-28-6000	1973	79	1200	.61	.64
L-1011	1970	268	6800	.44	.46
Concorde	1976	108	6230	3.3	3.3
B747-100	1970	374	6000	.42	.45
B707	1970	149	6000	.50	.55

Table 1 Prediction of Historical Direct Operating Costs

The indirect operating cost can be subdivided into two groups: aircraft related and non - aircraft related operating costs.

The aircraft related operating cost per km are:

Ground handling, equipment and landing fees are roughly proportional to the maximum takeoff weight, and inversely proportional to the aircraft range.

$$C_{o,10} = D_l \, 0.023 \, m_{to} / R \tag{24}$$

The navigation fees and the aircrafts administration per km are proportional to the maximum takeoff weight:

$$C_{o,11} = D_l \, (0.58 + 6.7\text{e-}7 \, m_{to}) \tag{25}$$

Cabin crew cost per km is proportional to the number of seats and inversely proportional to the cruise speed:

$$C_{o,12} = D_l \, 4.2 \, s / V_b \tag{26}$$

The non aircraft related operating cost are currently around 1.5 cts/s.km assuming a loadfactor l of 100 %. They are typically one time expenses per passengers and therefore inversely proportional to range. These costs include: passenger food, passenger service, baggage handling, reservation advertising, commissions and airline non-aircraft administration.

$$C_{o,13} = D_l \, 13 \, s \, l / R \tag{27}$$

$$\text{IOC} = \sum_{i=10}^{13} \frac{C_{o,i}}{s} \tag{28}$$

2.3.3 Operational Revenue - Passenger

The DOC is directly related to the cost of development and the cost of production to the manufacturer as well as the cost of operation. Therefore the DOC of two aircraft analyzed with the same method and conditions will allow both the airframe manufacturer and the airline to select the better aircraft[1]. This is not the whole story, however. One aircraft may be able to attract more payload than the other. Since passenger carrying aircraft should be compared at the same level of comfort only the time saved by the passenger would make him pay more for a ticket. In many publications, for instance by Mizuno and others [34], the additional revenue due to time savings was found by means of a questionaire. They found the value of time for economy travellers to be $40 (1990) per hour. The question is whether economy passengers would stay an hour longer in the airplane if they were paid less than $40 dollars an hour to do so. In the author's opinion more than half the passengers would. Based on Steiner's [35] book I propose to use the average hourly wage as the time value of the economy passenger. To compare one aircraft to a reference aircraft

---

1. This is not true if the pricing of the aircraft were determined by the market mechanism, rather than the actual production cost plus profit.

we have to subtract the additional revenue due to time savings of the aircraft with respect to the reference aircraft.

$$\Delta_{REV} = \frac{20 \cdot l \cdot D_l \cdot [T_{b,\,ref} - T_b]}{R} \tag{29}$$

In this expression $20 represents the estimate average hourly income of US air travellers in 1995. A corrected total operating cost for a balanced market can now be defined as:

$$TOC_{cor} = DOC + IOC - \Delta_{REV} \tag{30}$$

2.4 The Real Market

So far we have focussed on engineering tradeoffs in an equilibrium market where the manufacturer sells the required break-even production run and the airline sells the seats with the average load factor. Such a model may be very useful for preliminary economic evaluation of aircraft projects. In reality however, nobody wants to sell the same product at the same price as their competitors. In doing so they would invariably lower the price of the product offered.

The trick is to offer something at a lower cost and higher price. This can only be done by offering something new. As an example we will discuss the introduction of a supersonic aircraft under the assumption that it does not violate any existing performance or environmental regulations.

The manufacturer is now faced with two more variables: The number and price of the aircraft he can sell. Since he - we hope, but this is far from sure - knows what his production costs are, he would then have an idea of whether he could make a profit.

The aircraft number of units versus price is determined by a complex macro-economic model, that aircraft manufacturers such as Airbus can use as a subroutine in an overal aircraft optimization program.

The price and number of aircraft are determined by the iteration shown in Figure 8. Its starts with an estimate of the price and number of aircraft sold. We can now use the manufacturer and airline cost models to estimate the (future) total operating cost of the aircraft.

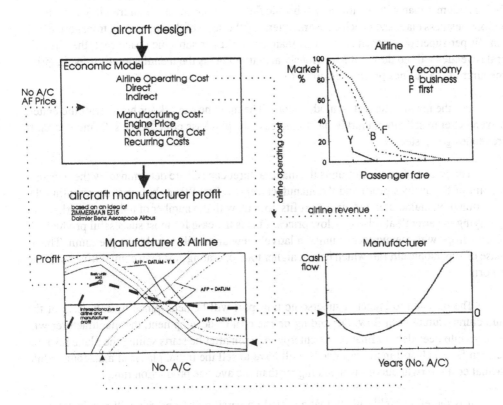

Figure 8   Simplified Macro Economic Model of Aircraft Units vs. Price

For a given aircraft service (speed) passengers are willing to pay various amounts of money on various routes. Passengers will be inclinded to pay more for shorter travel times. Typical business class passengers will also pay more than coach passengers, because their time is worth more. Figure 8 shows what fraction of passengers is willing to pay what as a function of the fare class. The airline will try to sell the seats as expensive as possible. So, as a function of the local route class distribution the airline will get more revenue. The airline has to sum all the revenue over all the routes an aircraft flies to decide whether to buy an individual aircraft.

This decision is made based on airline profit. The profit is the revenue minus the total operating cost. If the revenue of an individual aircraft purchase exceeds that of the aircraft it is replacing and there is no competing product, the manufacturer can expect to make a sale.

This 'mixed fleet' concept of supersonic and subsonic airliners is also shown in Figure 8 . The curves shown here are typical for supersonic transports which command higher seat prices at higher operating cost than current subsonic transports.

When very few aircraft are sold, the airline is doing well. It sells all the seats of the supersonic transport at a premium to first class passengers. The premium is higher than the premium they can charge in the more competitive subsonic markets, so the airline makes more profit with this fleet mix than with a uniformly subsonic fleet. As they buy more aircraft however, more and more business class and coach class travellers fill the seats of the supersonic transport. Since the profit per supersonic coach seat is less than the profit per subsonic coach seat, the airline's overal revenues will go down. It is obviously also true that as the manufacturer raises the price of the aircraft the airlines profit goes down too.

For the manufacturer it is another story. The more he can ask for his aircraft the better. It is even better to sell more aircraft than to increase the purchase price. As he sells more aircraft his revenues go up steeply.

The price and number of units the manufacturer can sell are determined by the intersection point of the airlines profit and the manufacturers profits. When the curve representing the intersection of manufacturer and airline profits goes up with the number of units, the market will keep buying the aircraft at relatively low prices. This is the case for most successful products. If the curve drops with the number of units, relatively few units will be sold at a premium. This is the case of supersonic aircraft which have a higher total operating cost than competitive subsonic transports.

The decision to launch a supersonic transport will be determined by estimates of the manufacturers future cash flow. Depending on the cost of development, the manufacturer will have to go into deep debt to finance the enterprise. Even after he starts selling the plane his negative cash flow will still increase since he will have to sell the initial planes at a discount while his actual cost - Wright curve - is much higher than the average production run.

It is therefore highly likely that a scaled up version of Concorde will suffer the same market fate as Concorde itself. Only when the total operating costs come down to values that are very close to those of current subsonic transports will the projected sales be large enough to justify launching such a project. International cooperation may reduce the cash flow risk for individual companies and ameliorate this situation somewhat.

The main problem with this macroeconomic model is that its inputs are determined by interviewing the custumer (airlines), and that his views may have changed by the time the aircraft is offered for sale. Therefore probability distributions have to be attributed to different scenarios.

2.5 Conclusions

The present work provides a tool to tradeoff commercial aircraft design features assuming a balanced market for subsonic and supersonic jet aircraft up to Mach 2.2. The model integrates older methods such as the ATA '67 and the American Airlines studies of the 1980's with newer data into one consistent whole.

Furthermore, we have presented a simplified qualitiative macroeconomic model that predicts aircraft price and the number of units sold. We showed that the simplest way to sell an aircraft is to offer a product with better or equal performance than the product it is replacing at the same price or to offer the same product performace at a lower price. Trading off higher performance at a higher price is highly speculative in an unknown market.

Acknowledgements

The author would like to thank Mr. Schwartz, Mr. Zimmermann and Mr. Kaeske of engineering cost and market forcasting at Daimler Benz Aerospace Airbus Hamburg for their input.

2.6 References

[27]   **A.I.A.A.**
       Aerospace Facts and Figures, 1972 - 1990

[28]   **American Airlines**
       A New Method for Estimating Current and Future Transport Aircraft Operating Economics, NASA CR-145190, 1978

[29]   **Anderson, J. L.**
       Price-Weight Relationships of General Aviation, Helicopters Transport Aircraft and Engines, SAWE Paper 1416, 1981

[30]   **Anderson, R.**
       Weight Estimation Methods, Unpublished notes Design Branch AFFDL Wright-Pat AFB, 1973

[31]   **A.T.A.**
       Standard Method of Estimating Comparative DOC's of Turbine Powered Transport Aircraft, December 1967

[32]   **Hartley, P.**
       The Learning Curve and its Application to the Aircraft Industry, *Journal of Industrial Economics*, March 1965

[33]   **Maddalon, D. V.**
       Estimating Airline Operating Costs, NASA CP 2036, 1978

[34]   **Mizumo, H. et al.**
       Feasibility Study on the Second Generation SST, AIAA-91-3104, 1991

[35]   **Steiner, H. M.**
       Public and Private Investments George Washington University, 1980

[36]   **U.S.D.T.**
       Aircraft Operating Cost and Performance Report, U.S. Department of transportation, Sept '85

[37]   **Van Ameyden, L. J.**
Evolution or Revolution in Aircraft Industry, Symposium Delft University of Technology, Sept 1985

[38]   **Van der Velden, A. J. M.**
An Economic Model for Evaluating High-Speed Aircraft Designs, NASA CR 177530, May 1989

[39]   **Wright, T. P.**
Factors Affecting the Cost of Airplanes, *Journal of Aeronautical sciences*, feb 1936

CHAPTER 3

SON OF CONCORDE, A TECHNOLOGY CHALLENGE

J. Mertens

Daimler-Benz Aerospace Airbus GmbH, Bremen, Germany

3.1 Concorde Technology Level

Concorde (Figure 9) is the only supersonic airliner which has been introduced into regular passenger service. It is still in service at British Airways and Air France without any flight accidents, and probably will stay in service for at least for ten more years.

Concorde has experienced the most supersonic flight hours and flight miles of any aircraft. Indeed, the twelve flying Concordes have accumulated more supersonic flight hours than the total of all military aircraft in the world. Concorde's range is about 6.500 km, whereas the best fighters like the Su-27 or F-22 (so called "supercruisers") achieve about 200 km in sustained low supersonic cruise, and military supersonic bombers or reconnaissance aircraft like the B-1 or SR-71 reach about 3.500 km without refuelling. But although the supersonic flight range of Concorde is by far better than for any other supersonic jet built, this range was the most important limiting factor in the commercial success of the Concorde. A new viable supersonic airliner, called Supersonic Commercial Transport (SCT), must be able to serve the important trans-Pacific market, requiring a range of 10.000 to 11.000 km. This is a tremendous improvement compared to the Concorde.

What are the differences between Concorde and a new SCT? Besides the larger size, which improves the range performance a bit, technology improvements are cited to enable this big step forward. So, let's look at the technology improvements we have achieved in aviation *since the Concorde design.* As there is no other supersonic airliner, we have to compare Concorde's contemporary subsonic airliners with the newest generation of subsonic airliners.

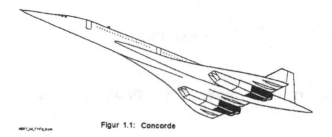

Figur 1.1: Concorde

Figure 9   Concorde

Aircraft *flight performance* is governed by aerodynamics, structures and engines. All other specific disciplines, although often important for the viability of an aircraft, are only weakly related to *flight performance* of transport aircraft. Therefore we will look at the improvements in these three main disciplines.

3.1.1 Concorde structure weight still "comparable" - 25 years later

The first example of technologies are weight improvements, listed in Table 2

	OWE %	PAY %	OWE / PAY	F / PAY·Mm	Improvement %	Range Mm	PAX No.	MTOW Mg
Concorde	42,52	4,8	8,90	**1,647**	---	6,58	98	185,1
B737-200 *)	56,14	18,4	3,06	**0,398**	---	4,07	107	52,6
B737-500 *)	52,01	16,1	3,23	**0,436**	-9,55	4,48	108	60,6
B747-100	48,90	10,5	4,67	**0,458**	---	9,04	385	332,1
B747-400	45,99	9,61	4,79	**0,345**	24,67	13,27	420	394,6
A340-300E	47,90	9,82	4,88	**0,315**	31,32 (rel. to B747-100)	13,24	295	271

Table 2 Weight improvements since Concorde

OWE: operating empty weight,   PAY: payload,   F: fuel
PAX: passengers   MTOW: maximum take-off weight
Part of improvement was used to increase range, part to increase payload
Performance comparison by fuel per passenger-kilometer
About the same development generation: Concorde, B737-200, B747-100
Modern aircraft, 25 years later: B737-500, B747-400, A340-300E
*)Data base for the different B737-versions seems to be inconsistent; because the more efficient engine of the B737-500 (CFM-56 instead of JT-8D for B737-200) should improve aircraft efficiency, at least for long ranges

Boeing B737-200 and B747-100 were developed in parallel with Concorde [40]. The new generation of comparable size are B737-500, B747-400 and Airbus A340-300E [41]; the latter being handicapped when compared with the B747-400, because it is a bit smaller. Improvements of fuel per passenger-kilometers of more than 30% were achieved; but comparison to the old aircraft is difficult, because data base has changed (improved seating standards etc.).

When comparing Concorde's structure weight, it is still comparable to the A340. (Comparison can only be made and was made by designing both aircraft with the same design tool. Assuming several structure technology standards, results showed the weights to reach the actual values of Concorde resp. A340, when using the same structure standards for both designs [42]). But what are the technologies leading to the subsonic weight improvement or Concorde's advanced values?

- Because Concorde has only a small payload fraction, it is much more sensitive to weight increments; therefore more effort was spent for weight savings and expensive solutions became useful.

- Many (weight) improvements for transonic aircraft after B737-100, B747-100 were provided by interdisciplinary effects like

  high bypass engines or

  optimized (nonlinear) transonic aerodynamics (Figure 10) via, increased wing profile thickness and volume, reduced wing sweep,

  which cannot be transferred to a Concorde type configuration.

- System weight was only marginally reduced since Concorde.

- Since Concorde's time many improvements in structure technology were not transferred into weight improvements, but offset by advanced safety requirements; e.g. new requirements for pressure losses, cabin evacuation or fire resistance of cabin equipment. But this is the world where a new SCT must fly.

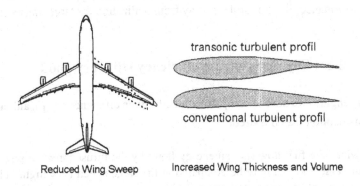

Figure 10   Improvements in Transonic Aerodynamics.

3.1.2 Limited aerodynamic improvement potential
for a Concorde-like aircraft

When comparing aerodynamic improvements since the Concorde's time, we find strong differences in the aerodynamic efficiency between Concorde time subsonic aircraft like B747 or B737 and modern aircraft like the B777, A340 or A320, expressed by improved L/D (lift/drag). These subsonic aircraft fly at high subsonic Mach numbers, when the air flow at the aircraft locally reaches partially subsonic, partially supersonic Mach numbers, which is named transonic flow. Physics of transonic flow include strong nonlinearities like shock waves, and the governing equations change from elliptic to hyperbolic type. Whereas the old aircraft were designed using pure subsonic linear potential flow theory combined with simple sweep theory, improvements were provided exploiting nonlinear theory; the latter require modern high performance comput-ers not available for Concorde development.

But flow around supersonic airliners like the Concorde is dominated by small disturbances of the incoming flow, because strong disturbances would create high wave drag. Therefore, design of supersonic airliners can mostly be based on linearized potential flow theory, as was the Concorde (slender body theory). Only for some parts nonlinear effects have to be respected for: strong interference effects like engine integration, fine tuning of the configuration and strongly nonlinear boundary layer flows. But, except for laminar flow, other strongly nonlinear boundary layer flows like separation are avoided because they are connected with large drag increases. Therefore, modern nonlinear aerodynamic theory can only provide limited improvements compared to Concorde (except for laminar flow).

Although, some aerodynamic improvements may be provided:

• New materials providing higher specific stiffness may allow a higher aspect ratio via interdisciplinary effects.

• Local optimization for nonlinear flow phenomena will provide reduced interference drag, especially for engine integration.

• Supersonic laminar flow is the only new aerodynamic technology which can strongly improve performance. But it is still far away from realization for large transport aircraft.

3.1.3 Olympus engine efficiency still very good

When comparing engines we have to compare installed engine efficiencies (η_E), although mostly specific fuel consumption (SFC) is used.

SFC divides the fuel flow (i.e. an energy flow) by the thrust force; so physically it is not a meaningful value and can only be compared at the same speed (v = flight velocity)). In contrast η_E is the amount of energy flow provided by engine thrust divided by the energy flow provided by the fuel (H = calorific value of fuel). η_E and SFC are connected via the equation

$$\eta_E \cdot H = v / SFC \tag{31}$$

When not using a consistent system of units like SI, the respective unit conversions have to be applied. For Kerosene H is given by

$$H = 42.817 \text{ MJ/kg} \tag{32}$$

Now we can compare engines. At Mach 2.0-cruise, Olympus [40] and a proposed new supersonic engine are given, as well as a modern high bypass transonic engine used for wide-body aircraft [41]:

$$\text{Olympus, M} = 2.0 \qquad SFC = 1.19 \text{ kg/daN/h} \qquad \eta_E = 0.41 \tag{33}$$

$$\text{New engine, M} = 2.0 \qquad SFC = 1.13 \text{ kg/daN/h} \qquad \eta_E = 0.43 \tag{34}$$

$$\text{CF6-80-C2, M} = 0.85 \qquad SFC = 0.56 \text{ kg/daN/h} \qquad \eta_E = 0.37 \tag{35}$$

We see that Olympus is still very good. The improvements of subsonic engines, mainly achieved by strongly increased bypass ratio, did not yet reach Olympus' efficiency. Indeed, at supersonic cruise a modern optimized engine with very low bypass ratio would provide slightly better values than the one given above. But this engine like Olympus will never meet the stringent noise criteria at take-off and landing which new SCT have to fulfill. Probably a bypass ratio of about 2 and extensive noise suppression (damping plus ejector) will be applied to meet noise criteria; this will decrease engine efficiency at supersonic cruise to the values indicated above.

3.2 Requirements for a new Supersonic Commercial Transport (SCT)

There are only a few Concordes, which serve a small, exclusive class of passengers over limited distances at high fares. For its rare presence Concorde is allowed to meet only elementary environmental criteria which will not become more stringent for Concorde itself. Especially for noise, ICAO Annex 16 and FAR 36 only require, that Concorde must not become even more noisy than it was at certification; and Concorde is very noisy. For new supersonic airliners those rules will never be applied; instead, a new SCT must fulfill new requirements which are not met by Concorde:

• It must comply with all valid certification rules,

- it must be economically viable,

- it must provide sufficient comfort.

A new SCT must be "just another aircraft" [43]. In the following paragraphs only those points are mentioned which will introduce significant new challenges compared to Concorde.

3.2.1 Relevant certification rules

Safety:

Concorde has proven to be safe. But since Concorde certification some new rules were introduced by the authorities e.g. FAA, JAA, which a new SCT has to meet. The most obvious challenge is cabin pressure loss.

It is required that aircraft and passengers can survive pressure loss in the cabin provoked by a sudden hole like a broken window. Therefore, sufficient pressure levels must be maintained in the aircraft when a hole opens. After DC10 accidents, when a burst of underfloor cargo doors destroyed the floor of the aircraft with the hydraulic systems, size of this hole was increased by pure geometrical definition (about door size); for a wide body aircraft it has a size of 20 sqft (FAR 25). This poses extreme difficulties for flight altitudes above 40000 ft. It is impossible to fight against this requirement by building a stronger aircraft, because there is no requirement on strength or probability of creation of such a hole, but only a given hole size. Concorde needs only to survive a hole in the hull of window size; and Concorde has small windows.

Environment:

- Concorde produces unacceptable noise at take-off and landing, although having only 110 passengers. A new SCT, which will be of at least double the size of Concorde, has to meet current noise standards (FAR 36, stage 3) or even more stringent ones. For comparison: FAR 36, stage 3 is just met by the B747-200. But SCT engines will not have a bypass ratio of about 5 like the B747-200 engines [40], but only one of about 2.

- Best aerodynamic performance L/D (lift/drag) is reached at elevated values of the lift coefficient (about $C_L=0.5$ for subsonic aircraft, below $C_L=0.15$ for supersonic flight). To fly at good aerodynamic performance, to maintain acceptable pressure levels in the inlets and engines, and to fill the engines with air, dynamic pressure at supersonic cruise will be in the range of about 20 to 30 kPa. Therefore, flight altitude depends on cruise speed and weight, for a Mach 2 aircraft about 16 km. And the higher the cruise altitude, the more sensitive is the atmosphere to pollution, especially the ozonelayer. At the time being it seems that a Mach 2-SCT will not harm the ozonelayer; but this is based on calculations which are still questionable. Supersonic aircraft will burn more fuel per passenger kilometer than subsonic transports. Although CO_2 is not altitude sensitive, it is a well known greenhouse gas, and the large amount of CO_2 emitted has to be justified. In the future it has to be expected that the public will become even more sensitive to environmental impacts. Therefore a new SCT has to demonstrate, that its impact on the earth's atmosphere is tolerable.

- A body cruising at supersonic speed generates a sonic boom which follows the body. This boom is an annoying and startling noise in an area of about 20 to 40 km at both sides of the SCT's track. To avoid harassment or damage, civil supersonic flight will only be permitted over sea or perhaps uninhabited land. Because noise of the natural environment is so high on the sea, there is no harassment or damage known to people, animals or ships below Concorde routes. In contrast to people on land, there exists no complaint about Concorde's sonic boom over sea [44].

Operations:

- Any new aircraft must be able to follow current and future ATC (air traffic control) procedures. In the future, steeper descent angles may be requested in the airport area, which could be a challenge for SCT.

- New SCT must meet current ground load values, i.e. wheel number, size, loading and distribution. High ground load values may damage some airports, especially on aprons with tunnels.

- Loads on passenger and crew during operation must stay within acceptable boundaries. Especially long elastic fuselages provide strong vibrations during ground roll and in turbulence, and may impose high g-loads during rotation.

3.2.2 Economic viability

Range:

Concorde is only able to serve some of the medium range overwater routes. This is an inattractively limited part of the market. To gain a sufficient part of the market, future SCT must serve the important overwater long range routes. This, for the trans-Pacific routes, is more than 5500 nm range. Range is even more important than speed! For its limited range Concorde did not find a market, although Concorde's range was at the achievable limits. Also, the American SST, the Boeing B2707, failed for its insufficient range [45] and its uncontrolled growth of weight, even when environmental concerns were cited to stop the program [46] (which luckily limited the liability of Boeing).

Operations:

Today, there exists a very expensive environment for air traffic operations: airports with its buildings and installations, hangars and air traffic control (ATC). New aircraft may require some additional installations, but the necessary investments must be paid by the money earned with those aircraft. And at least for the introduction of the new aircraft, success depends on the ability to cope with the existing environment. Time spent at the airport for debarking and embarking of passengers, servicing of the aircraft, refuelling etc. must be minimized:

- Aircraft dimensions (length, span or occupied ground area) must be compatible with existing installations at the relevant airports.

- Aircraft accessibility (doors) must be compatible with existing airport installations; service ports should be compatible with usual procedures.

- Aircraft accessibility must allow for parallel de-/embark, service, fueling, etc.

- Aircraft supply needs must be compatible with existing (big) systems, e.g. fuel type.

- To serve the many routes with overland legs, subsonic cruise performance must be about as good as the supersonic one. Concorde has only poor subsonic performance.

Cost and fares:

A new SCT will become a success only if the manufacturer of the aircraft and the airlines operating this aircraft will earn money with this aircraft. This requires production of a sufficient number of aircraft at a profitable price for the manufacturer, and operating costs which are lower than the money paid by the passengers:

- To reduce seat mile cost and to serve a large market, an SCT must transport many passengers (size effect), at least as much as subsonic long range aircraft (e.g. A340). (Smaller SCT would -at first- only serve full fare passengers; but after a short period airlines will begin to introduce reduced fares to fill empty seats -because empty seats are the most expensive seats-, and this leads to the situation we have today with mostly low fare tickets.)

- To gain a sufficient part of the market, a future SCT must serve all classes.

- Speed pays (see cars, trains, air traffic) but a surcharge must stay in acceptable limits. It seems that the travelling public accepts surcharges of about the money they would earn for the time they saved by increased speed, even for holiday trips.

- An SCT with those (low) fares must still allow a profit for the manufacturer and airlines. And the manufacturer must sell many aircraft in order to provide an acceptable price. This becomes impossible with pure exclusivity.

3.2.3 Comfort

Passengers like comfort, and especially on long trips some level of comfort is necessary. But what is the special kind of comfort required for supersonic transports?

Range:

Passengers select a faster transport to save time; and they pay for the time saved. But this is the time from airport to airport or even house to house. Therefore supersonic transport only makes sense when the flight time is an essential part of the trip time, i.e. for long ranges.

And passengers do not like stop-overs, because they prolong flight time and are even more annoying than flight.

Therefore an SCT must provide long range capability to connect at least the most important areas of the world: Europe, US and eastern Asia, all separated by up to 5000 to 6000 nm.

Speed:

> Passengers pay for the time saved. Therefore overall trip time has to be reduced, as well by high cruise speed without stop-overs, but, as importantly by accelerated check-in / check-out.

> Additionally, many passengers feel very uncomfortable during and after trips of more than 6 hours flight duration, especially children, or disabled and elderly people, who will feel circulatory trouble.

Space:

> Passengers want space similar to comparable subsonic transports, i.e. medium range transports with same flight duration. The narrow Concorde fuselage is only accepted for Concorde's exclusivity.

3.3 Estimation of Technology Influence

There are several technologies to improve a new SCT. In order to evaluate the importance of a technology, we must be able to estimate its influence on the realisation of an SCT. At present level of knowledge, there is a hierarchy of technologies from fulfillment of constraints over cruise performance to operations. The operations are still at the last position, because we are on search for a technical solution. When a solution is found, operations will become more important, because they contribute strongly to cost performance.

3.3.1 Technologies to fulfill constraints

Here technology estimation is possible by looking for the relevant physical principles. Influence on aircraft cruise performance and cost is indirect, but can be very strong and limiting.

The most important constraints for an SCT are:

Take-off:

• Field length:

> The required *thrust* (i.e. engine size) is determined by *weight* (acceleration) and *aerodynamic lift* (span, wing area, rotation angle) at lift-off (i.e. minimum airspeed).

> Take-off field length is defined by runway length of the relevant airports to be used by SCT. Today about 11000 ft are assumed.

• Climb rate:

> The required *thrust* is determined by *weight* and aerodynamic performance *L/D*.

> After take-off, a climb rate with one engine out must be maintained of

0,5% with gear extended and
3% with gear retracted.

Noise:

Noise is determined mainly by *exhaust velocity* of the jet engines (as long as the turbomachinery is well shielded by inlet and nozzle) and mass flow (i.e. *thrust*) (take-off thrust = massflow · exhaust velocity).

Stage 3 compliant low noise exhaust velocities are between 300 m/s (Airbus A340, well below stage 3 limits) and at most 400 m/s [47]; for compliance, bypass ratios of about 2 or comparable measures are required.

Low speed *thrust* is determined by field *acceleration*, by *weight* and *drag* during climb, and by *drag* during approach.

In the rules of ICAO, annex 16, chapter 3 and its derivations FAR 36, stage 3 or JAR etc., maximum noise allowed at take-off and landing is defined. Figure 11 shows the three points, where noise is measured. Maximum noise levels allowed for the different points depend on aircraft weight and number of engines. Noise levels are measured in EPNdB, which is a time integral of the EPNL(dB) weighted noise. Additionally, if noise at one point is only a little more elevated (at most 2 dB), it may be compensated by lower noise at the other points, following some complicated weighting.

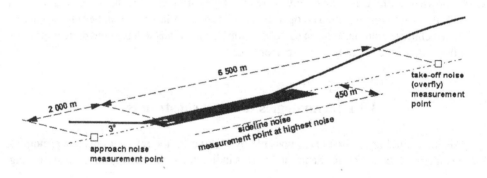

Figure 11   Noise Measurement Points

Transonic acceleration:

The highest *wave drag* values and the worst *engine efficiency* at nearly the same low supersonic speed may determine engine size.

Best supersonic cruise performance does not make any sense, if the supersonic aircraft is not able to reach supersonic speed.

Range:

Fuel amounts to a very high portion of gross weight (50% or even more). Therefore *cruise efficiency* determines range capability.

Airlines indicate, that a viable supersonic transport must be able to fly at least about 5 500 nm. Many routes include segments over inhabited land, where only subsonic speeds are allowed.

Supersonic cruise:

Reduce *aerodynamic drag* (here by slenderness), reduce *weight* and improve supersonic *engine efficiency* (here via decreased bypass ratio), to meet range requirements.

Drag at supersonic cruise is dominated by wave drag, but friction drag and induced drag are important as well. Drag has to be balanced by thrust.

Subsonic cruise:

Improve subsonic *engine efficiency* (here via increased bypass ratio) and reduce *aerodynamic drag* (here mainly larger span); both reduce supersonic efficiency.

Drag at subsonic cruise is dominated by friction drag and induced drag, possibly vortex drag due to separation. Drag has to be balanced by thrust.

Controllability:

Provide *control authority* for all disturbances either external (gusts, manoeuvers) or internal (failure cases like inoperative engines, cabin pressure loss).

Emissions:

It seems that most important will be NO_x-*generation* for its influence on the ozonelayer.

NO_x-generation is determined by peak temperatures which occur only at spot points in the burner, where stochiometric conditions are met. Low NO_x-burners reduce peak temperatures, but still maintain high mean temperatures.

Some of those constraining influences work against each other. For estimation of values compare subsonic aircraft of similar weight, e.g. B747, but pay attention to the significant physical parameters.

3.3.2 Technologies to improve cruise performance

To estimate influences on cruise performance, at first we will estimate the most important physics of constant cruise. This is, for a real aircraft, some kind of oversimplification, because it omits the important segments of take-off, climb, acceleration and reserves for go-around and divert as well as the minor parts for descend, landing and taxi. But for a long range aircraft it remains the most important segment.

Constant cruise is governed by the two equilibriums of lift (L) and weight (W), and of thrust (T) and drag (D):

$$L = W \qquad\qquad (36)$$

$$T = D \tag{37}$$

Thrust is paid by fuel (F)

$$T \cdot SFC = dF/dt \tag{38}$$

(SFC: specific fuel consumption, see eq. (31); t: time) which reduces mass (m) or weight

$$dF = - dm, \quad W = m \cdot g \tag{39}$$

but provides range (r) via speed (v)

$$dr = v \cdot dt \tag{40}$$

When introducing the other equations in (38), simple algebraic operations provide the differential Breguet-equation

$$dr = -\frac{v \cdot \dfrac{L}{D}}{SFC \cdot g} \cdot \frac{dm}{m} \tag{41}$$

or with engine efficiency η_E and calorific value of fuel H instead of specific fuel consumption SFC (eq. (31))

$$dr = -\frac{L}{D} \cdot \eta_E \cdot \frac{H}{g} \cdot \frac{dm}{m} \tag{42}$$

Simple integration gives Breguet's range formula

$$r = \frac{L}{D} \cdot \eta_E \cdot \frac{H}{g} \cdot \ln\left[\frac{m_0}{m_0 - F}\right] \tag{43}$$

or

$$r = \frac{v \cdot \dfrac{L}{D}}{SFC \cdot g} \cdot \ln\left[\frac{m_0}{m_0 - F}\right] \tag{44}$$

Aerodynamics determine aerodynamic performance L/D and so influence needed fuel F, but also provide requirements for structural layout and systems as well. Fuel, structures, engines, equipment and payload are mass m_0. Engines contribute to mass and provide its efficiency, η_E, which influences needed fuel, F and mass for fuel storage. The selected cruise speed

v or cruise Mach M influences achievable L/D, engine efficiency η_E or SFC, needed fuel F, and structure and equipment mass.

So, the Breguetequation allows simple estimation of range, but it is still difficult to estimate improvements by singular technologies. But the Breguetequation provides a rough estimation of range improvements, when only one parameter is changed, e.g. range improvement Δr provided by a pure aerodynamic improvement $\Delta L/D$ for the same aircraft (η_E, m_0, F). Now equation (43) gives

$$\Delta r = \Delta \frac{L}{D} \cdot \eta_E \cdot \frac{H}{g} \cdot \ln \left[\frac{m_0}{m_0 - F} \right] \tag{45}$$

For fixed range, we can try to estimate the influences on fuel needed, (F), by the inverse Breguet-equation

$$\frac{m_0}{m_0 - F} = \exp^{\left[\frac{r}{\frac{L}{D} \cdot \eta_E} \cdot \frac{g}{H} \right]} \tag{46}$$

But caution: now both m_0 and F usually become strongly dependent on $\Delta L/D$ or $\Delta \eta_E$ because better drag or engine efficiency reduces fuel consumption F, and this reduces gross weight m_0. Therefore the inverse Breguetequation usually calculates an aircraft whose performance is not exploited. To exploit performance, the aircraft has to be resized. This cannot be calculated by a simple equation. As a first, very rough estimation, the unresized improvement for fuel consumption, calculated by the inverse Breguet-equation (46), has to be multiplied by the factor:

$$1 + \text{(weight part under consideration)} / m_0 \tag{47}$$

e.g. for improvement of η_E: $(1 + F/m_0)$. But remember; The Breguetequation only calculates idealised cruise flight. Therefore it may only be used for estimation of tendencies in long range flight.

Small improvements in fuel consumption by individual technologies may be *compared* by the percentage of the individual improvement, multiplied by

$$\text{weight part influenced by this technology} / \text{payload.} \tag{48}$$

This is explained in Figure 12 by several examples for an Airbus A340-300B with an assumed 15% pure drag improvement or a 15% structure weight reduction; for all examples the maximum take-off weight (MTOW) was maintained by adding fuel or payload.

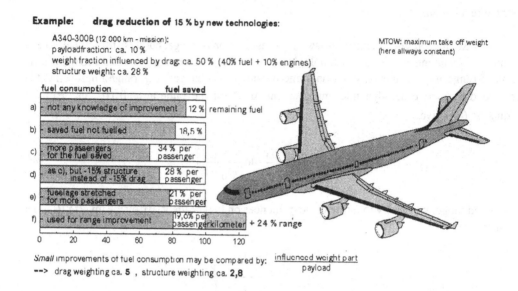

Figure 12   Improvement estimation

3.3.3 Technologies to improve operations

Technologies for operational improvements usually are not directly connected to flight perform-ance, but rather to operating costs. Therefore, simple estimation of their influence is difficult. Most of these technologies are not specifically related to SCT. Here, only some of them are mentioned.

Materials with improved creep resistance for hot engine parts:

> In supersonic engines temperatures remain high during cruise, whereas for subsonic air-craft peak temperatures are reached only at take-off. Therefore high temperature creep re-sistance becomes more important for SCT. It will allow higher engine (cruise) efficiency and/or reduce maintenance costs.

Special control systems to guaranty controllability:

> Because SCT-configurations strongly differ from subsonic or military aircraft, some dif-ferent control problems may occur; e.g.

>> • inoperative engines with high lever arms in supersonic flight; if this determines control surface loads (mainly rudder, but also aileron, elevator) special surfaces may provide improved solutions

>> • after a sudden pressure loss

>> • special devices or differently sized devices to produce high drag may enable rapid

descent and deceleration to reach sufficient atmospheric pressure levels

- special devices may help to maintain sufficient pressure and temperature in the cabin.

If needed, appropriate systems must be developed.

Neutral point and center of pressure vary strongly when going from take-off over transonic flight to supersonic cruise and back for landing. For optimum flight performance during supersonic cruise, only very small control flap deflections are allowed.

Both require a highly sophisticated fuel trim system.

Artificial vision will avoid a Concorde-like droop nose.

Artificial vision is just under development for CAT III landings and ground roll of subsonic aircraft.

Manoeuver and gust load alleviation may reduce wing weight.

Just in use for several aircraft (L1011-500, A320, A330/A340) and will be improved for future ones (A3XX).

Systems for reduced turn-around time improve aircraft productivity.

Such systems are under development all the time, especially for expensive, very large aircraft like A3XX. Special problems for SCT are

- a large wing root blocks accessibility to large parts of the fuselage

- a narrow fuselage retards boarding and ground cabin service (like cleaning)

- the large fuel amount requires several points or long time for fuelling.

Systems to reduce maintenance costs are under development for subsonic aircraft all the time. Specialities for SCT are

- supersonic inlet and nozzle are specific to SCT

- accessibility to the engines is reduced by supersonic inlet and nozzle, and possibly by installation just at the wing

- in many areas of the SCT space is very limited (thickness at movables, tail, nose)

- dissimilar SCT-geometry requires different procedures

- elevated temperatures during cruise introduce dissimilar loads and load cycles which alter maintenance, sometimes even in favor for SCT (e.g. corrosion).

Active landing gear allows better damping of roll vibrations, ground loads and supports take-off rotation.

Just under development for very large aircraft like A3XX.

Improved ATC-systems

Strongly required are new ATC systems and procedures, especially for long range overwater guidance and area navigation (direct flight, not restricted to airways linking ATC-control points), and automated data links. They are just under development for subsonic aircraft.

3.4 Technologies for a Concorde Successor: Kind and Risks

We have seen, that many technologies required for a new SCT are just developed or are under development for subsonic aircraft. Other technology improvements cannot be transferred to an SCT. And Concorde's technology level is still "comparable". But what are the new technologies which will enable realization of a competitive SCT with sufficient range and operating costs?

3.4.1 Aerodynamic technologies

There are several chapters in this book dealing with aerodynamics. Therefore, here only the important points are mentioned.

Aerodynamic multi point design:

For subsonic aircraft, in the first design step usually a two point design is made for

- high speed cruise and

- low speed take-off and landing.

Often this is achieved via a nearly pure one point design for high speed cruise, whereas low speed performance is achieved using rather complicated flap systems.

A supersonic aircraft must be able to cruise economically at supersonic speed over uninhabited areas and at high subsonic speed over inhabited land. And supersonic cruise only becomes possible if transonic acceleration can be realized. So, we have to design for four points:

- supersonic cruise,

- high subsonic cruise,

- low speed take-off and landing,

- transonic acceleration.

Moreover, low speed performance cannot be achieved using additional large flap surfaces like fowlers. Since an SCT has a large wing surface, effect of (fowler) flaps is very limited. And lift slope ($C_{L\alpha}$) decreases for small aspect ratio. But lift via high angle of attack, as used for Concorde, is accompanied by large drag, which requires noise producing thrust.

Therefore a best combination of aerodynamic's (and other discipline's) design principles is required to meet the diverging requirements. The most important part is selection of a suited configuration, for which aerodynamics contribute strongly in meeting the four aerodynamic design points.

Nonlinear aerodynamics:

Nonlinear aerodynamics enable minimization of interference losses, e.g. engine and wing-fuselage integration. It is being developed for subsonic transports; but it becomes more challenging for supersonic transports by the combination of nonlinear effects with very weak oblique shocks, strongly three dimensional geometries and boundary layers, and strong shocks around the engines and in the inlets, even with reflected shocks.

Control surfaces:

Design of the control surfaces depends strongly on the configuration selected. Smart solutions may decide on the viability of a configuration or strongly ease design and operation.

Aeroelastics:

New approaches like aeroelastic tailoring or aeroelastic control are being developed for large subsonic transports. For the slender SCT-configuration, inclusion of aeroelastics in the early design is a prerequisite, at least for static aeroelastics (i.e. inclusion of the shape variation due to aerodynamic loads, but still without vibrations). For many SCT-configurations, flutter itself is as important, but hitherto it can be checked only at a more matured design stage, when a more detailed data base was built up.

Aerodynamics, especially nonstationary aerodynamics and dynamic structure calculations are much more challenging for an SCT than for subsonic transports, at least for the flutter sensitive symmetric (Concorde-like) configurations. Because flow and structure dimensionally has to be treated completely three dimensional by 2D- or quasi-2D approximations like airfoil flow, sweep theory or beam approximations are not possible.

Aerodynamic damping decreases with speed. Therefore supersonic cruise becomes flutter sensitive. But the nonlinear transonic aerodynamics (here high subsonic) also decrease flutter damping, the so called transonic dip. So there is a second high subsonic flutter case which often is even more flutter critical than supersonic cruise [48]. This is one of the most challenging calculations in SCT design.

Supersonic laminar flow (SSLF):

Supersonic laminar flow (SSLF) may provide strong improvements during cruise flight. But it is still far away from realisation and many questions are unresolved; even some physical principles are still not yet understood.

Today, we postulate a reference SCT to be viable without SSLF. Because, when a new subsonic aircraft benefits from laminar flow, the SCT must also be improved by SSLF in order to maintain its competitiveness.

3.4.2 Material technologies

An SCT has only a small payload fraction; therefore it becomes very weight sensitive. To make a Concorde-like SCT viable, the necessary reduction of structure weight (compared to available technology, e.g. A340) will be about

> 30% for a Mach 1.6 SCT
> 40% for a Mach 2.0 SCT
> 50% for a Mach 2.4 SCT.

Even when considering for the high unit price of an SCT, this seems to be very risky! But still some improvements in other technologies, especially configuration selection, may reduce the weight improvement requirements given above.

New Materials:

For Mach numbers below 1.8, highest temperatures occur in sunshine on the ground. For Mach numbers above 1.8, cruise temperature becomes important, especially in respect to life time. Below Mach 2.0, emphasis for the airframe is still more on light weight than on temperature. Some materials envisaged are carbon fibers (CFRP), metal matrix composites (MMC) for highly loaded parts and ceramics.

For supersonic engines creep resistant high temperature materials are required (peak temperature at cruise).

Manufacture:

New methods are required to manufacture very light weight structures and elements. For new materials, methods must be developed to fabricate parts and to join them to assemblies.

Concepts for inspection, repair and crashworthiness are required.

All these have to be qualified early.

Structure design:

Materials and manufacturing principles must be integrated in the design process. For higher Mach numbers, the design must consider for thermal dilatation and stress.

Optimization methods will reduce weight and integrate structure design with other discipline's needs.

Aeroelastics:

Provide data early in the design process, to direct the interdisciplinary design to an optimum solution which considers for aeroelastic deformation and which will have a safe flutter margin.

Integrate stiffness (and thermal) design with aerodynamic design (aeroelastic/ aerothermoelastic tailoring).

Certification:

New materials and manufacturing methods must be certified, before being considered in design.

For Mach numbers above 1.8, thermal fatigue properties must be demonstrated for the materials, for structural concepts, and for the aircraft itself.

3.4.3 Engine technologies

Engines must fulfill several requirements at different design points, like the aircraft.

Efficiency:

SCT-engines must be very efficient at supersonic and high subsonic cruise, and must provide sufficient thrust at take-off, transonic (low supersonic) acceleration and cruise.

Emissions:

To protect the atmosphere, pollution has to be minimized. Especially low NO_x-burners must be developed. This is even more stringent than for subsonic aircraft, because SCT fly higher and burn at higher temperatures during cruise.

Noise:

Engine noise during take-off and landing must be comparable to subsonic aircraft (ICAO annex 16, chapter 3; FAR 36, stage 3). This requires large nozzle exit areas, comparable to subsonic aircraft of the same weight.

The multipoint capabilities of the engines are strongly related to the configuration selected. For example thrust available during take-off and landing depends on engine bypass ratio, noise reduction by suppression or simply throttling down, possible integration of noise suppressors in the airframe etc. This influences engine weight, and thrust available and engine efficiency in the other design points. On the other hand, thrust required in the design points is determined by the configuration, mainly aerodynamic performance and weight.

3.4.4 System technologies

Most systems are comparable with the ones of subsonic aircrafts. Emphasis is on low weight, small space and possibly high temperatures. Special systems for SCT are:

Inlet and nozzle control

Supersonic engines work at subsonic speeds. Inlet must decelerate the flow to subsonic speed, the nozzle must adapt to free stream pressure. Especially inlet control with several shocks and tuning with nozzle state is especially challenging. Concorde's solution is still state-of-the-art.

CG-control

From subsonic to supersonic flight, the center of pressure varies strongly. Control is provided by center of gravity (CG) control via fuel transfer. This system is proved in Concorde and applied in several subsonic jets as well.

Scheduled systems

In order to meet noise requirements at take-off and landing, it is envisaged to use an automated system for scheduling of flight path, flap settings and engine controls. This is envisaged for subsonic aircraft as well. But for an SCT it will be more complicated, and an SCT will be more dependent on such a system.

3.4.5 Multidisciplinary Design Optimization (MDO)

A "classical" Concorde-like design will be required to reach technology limits of all relevant disciplines. This requires perfect harmonization of all aspects in the design.

An unconventional configuration can improve performance over the limits of the "classical" configuration. But it requires a new kind of cooperation between the individual disciplines, with dissimilar interfaces between the disciplines compared to the "classical" approach.

In both cases, a solution will only be reached by using the tools of Multidisciplinary Design Optimization (MDO) which is treated in more detail in several lectures of this course. There are three objectives of MDO, each of them of equal importance.

Harmonize multiple disciplines:

Organize cooperation and data transfer among all relevant disciplines.

Although all over the world companies talk about introducing MDO, total quality management (TQM), concurrent engineering (CE) etc., in real life there are strong objections of hierarchies against any kind of cooperation between departments. The future will show if market competition will improve the situation.

Cooperative design:

Data to be respected for harmonization have to be carefully selected and must be ordered from crude predesign to detailed final design. Data transfer from pre design to the more detailed design steps and vice versa must be organized and fit into the data selected for interdisciplinary transfer.

Combine the relevant data into a design which must be evaluated by all participating disciplines.

The design process must be able to update the design exploiting the data corrections occurring in the cooperative evaluation.

Optimization:

Use optimization tools which are well adapted to the different design stages in order to exploit the best combination of the available technologies. Especially for interdisciplinary work, numerical optimization is recommended.

MDO is the *key technology* for a new supersonic transport. For subsonic transports remarkable improvements are also expected.

3.5 References

[40] **Jane's**
      All the World's Aircraft, 1977/78

[41] **Jane's**
      All the World's Aircraft, 1995/96

[42] **Van der Velden, A., von Reith, D.**
      Multi-Disciplinary SCT Design at Deutsche Aerospace Airbus. Proceedings of the 7th
      European Aerospace Conference EAC'94 "The Supersonic Transport of Second Genera-
      tion", Toulouse, 25-27 October 1994, paper 3.61

[43] **Frantzen, C.**
      Introduction to Regulatory Aspects of Supersonic Transports. Proceedings of the Euro-
      pean Symposium on Future Supersonic Hypersonic Transportation Systems, Strasbourg,
      November 6-8, 1989, paper II, 3.1

[44] **Mertens, J.**
      Sonic Boom Overwater Issues and Past Test Data, Review and Recommendations
      DA-Report DA-010-93 / EF-1971, Bremen, 6.10.1993

[45] **Goldring, M.**
      A Second Generation Supersonic Transport, the Lessons from Concorde. Proceedings of
      the 7th European Aerospace Conference EAC'94 "The Supersonic Transport of Second
      Generation", Toulouse, 25-27 October 1994, paper 2.31

[46] **Swihart, J.M.**
      Prospects for a Second Generation High Speed Civil Transport. Proceedings of the 7th
      European Aerospace Conference EAC'94 "The Supersonic Transport of Second Genera-
      tion", Toulouse, 25-27 October 1994, paper 2.33

[47] **Michel, U.**
      How to Satisfy the Takeoff Noise Requirements for a Supersonic Transport. AIAA-paper
      AIAA-87-2726, AIAA 11th Aeroacoustics Conference, Oct. 19-21, 1987, Palo Alto,
      CA, USA

[48] **Barreau, R., Renard, T.(Reporters)**
      BRITE EURAM Program "Supersonic Flow Phenomena", Final Report Subtask 1.3
      "Preliminary Aeroelastic Investigation of Supersonic Transport Aircraft Configuration"

CHAPTER 4

AERODYNAMIC MULTI POINT DESIGN CHALLENGE

J. Mertens

Daimler-Benz Aerospace Airbus GmbH, Bremen, Germany

4.1 Introduction

In the chapter "Son of Concorde, a Technology Challenge" one of the new challenges for a Supersonic Commercial Transport (SCT) is multi-point design for the four main design points

- supersonic cruise,

- transonic cruise,

- take-off and landing,

- transonic acceleration.

Besides engine technology, aerodynamics is most challenged by these differing requirements. But aerodynamic solutions will only become viable when contributing to an optimum of the whole aircraft, which is to be found in cooperation with all disciplines. Here, we deal only with the most important aerodynamic parameters at the different design points and consequences for aerodynamic design.

4.2 Supersonic Cruise

Supersonic cruise is the longest cruise leg on those routes which are of interest for supersonic transport. So supersonic cruise performance largely determines range, weight and cost of an SCT. Supersonic cruise performance strongly relies on aerodynamic drag reduction. Drag related weights (i.e. fuel and engine size) contribute about 60% to the SCT's MTOW (maximum take-off weight).

Here, we concentrate on drag reduction. Although, it has to be remembered, that drag minimization for itself is no goal for an aircraft: aerodynamic performance is L/D which is Lift divided by Drag or weight divided by drag; but, when looking for performance, more weight requires an equivalent increase in fuel consumption, see e.g. Breguet's formula, eqs. (44) to (47). A careful balance must be achieved when drag reduction requires a weight increase.

To estimate the influence of individual contributions to supersonic cruise drag, we look at the composition of (supersonic) drag. Contributors are skin friction drag, volume wave drag, lift dependent wave drag and induced drag. This is for the drag coefficient C_D

$$C_D = A/S \cdot C_f + K_v \cdot 128 V^2/(\pi S l_o^4) + K_L \cdot \beta^2 S C_L^2/(2\pi l^2) + K_I S C_L^2/(\pi b^2) \quad (49)$$

or in physical units for the drag D

$$D = A C_f q + K_v \cdot 128 V^2/(\pi l_o^4) \cdot q + K_L \cdot \beta^2 W^2/(2\pi q l^2) + K_I \cdot W^2/(\pi q b^2) \quad (50)$$

with

A:	total surface
b:	span
C_D:	drag coefficient
C_f:	friction coefficient
C_L:	lift coefficient
D:	drag
K_I:	shape parameter for induced drag (Osswald factor)
K_L:	shape parameter for lift dependent wave drag
K_V:	shape parameter for volume wave drag
l:	lifting length
l_o:	total (aircraft) length
M:	Mach number
q:	dynamic pressure
S:	wing (reference) area
V:	volume (aircraft + engine stream tube variation)
W:	weight

and

$$\beta^2 = M^2 - 1 \qquad\qquad (51)$$

In the formulas (49) and (50) some parts are bold faced. These are the contributors which can significantly be influenced by design.

Skin friction drag:

The first term describes friction drag which is mainly determined by the aircraft's total surface A and the friction coefficient C_f. For minimization both must be minimized.

Total aircraft surface A depends on the specified payload and the required fuel volume, aerodynamic lift, the slenderness and span requirements of other drag part minimization and (weakly) the aircraft's mechanical strength and stiffness. The total surface minimization is no goal for its own, it is rather only one aspect of drag minimization to be balanced with other requirements.

Friction coefficient C_f here is the mean value of the local friction coefficients, weighted by surface area. It is mainly determined by the local Reynolds-numbers. But it may be reduced by drag reduction techniques; those which are most well known are laminarization and riblets.

Laminarization maintains the low drag of laminar flow in the boundary layer, which reduces local C_f-values for an SCT by about 90%, but requires a huge technology effort (see special lecture). Presently, we design our SCTs without laminar flow, although laminar flow promises strong improvements. But an SCT which will become viable *only* with application of laminar flow, may loose its competitiveness with subsonic aircraft as soon as the subsonic aircraft utilizes laminar flow. Most likely, subsonic aircraft will apply laminar flow earlier than SCTs. On the other hand, an SCT concept which is viable without laminar flow, may be improved by laminar flow just like the competing subsonic aircraft.

Riblets are streamwise microscopic valleys in the aircraft's surface which reduce aircraft turbulent drag locally by about 8% [49].

Other measures to reduce skin friction may arise in the future. Special surface coatings [50] such as some nano materials (thickness of only a few molecules) have been proposed which alter the wall condition ("no slip") for the boundary layer flow, e.g. by not completely diffuse reflection of the air molecules at the wall.

Volume wave drag:

The second term describes volume wave drag which is mainly determined by the aircraft's volume V and strongly by slenderness V/l_o2, because l_o is squared and the whole term $(V/l_o2)^2$ is squared once more.

Total aircraft volume V is, like surface A, given with the specified payload and the required fuel volume, the aircraft's mechanical strength and stiffness and the slenderness and span requirements. Due to the important volume wave drag contribution, volume of

supersonic aircraft must be minimized, see Concorde's narrow and uncomfortable fuselage.

Larger total aircraft length l_o decreases volume wave drag and decreases structure
strength as well so that it increases aircraft weight and reduces stiffness. Both have to be
kept in acceptable boundaries. Especially for large SCTs, i.e. long fuselages, fuselage
flexibility poses big problems.
After Concorde's first landing at the old airport in Singapore, the pilots -although carefully fixed with seat belts in the tiny cockpit- hit the overhead panels with their heads due to
strong bending oscillations of the slender fuselage on the rough runway. This was, for a
long time, the last Concorde landing in Singapore.
Today airports request a maximum fuselage length of about 80 m, although most large
SCT-designs (250 or more passengers) have fuselages of more than 90 m which is at the
flexibility and weight limits.

The shape parameter K_V depends on the volume distribution of the whole aircraft including cross sectional variations of the engine stream tubes. At M=1 this is simply the distribution of the area normal to the flow direction (because Mach angle μ is 90°). At higher
Mach numbers it becomes more complicated:
From each point of the aircraft's surface a perturbation wave is radiated downstream
along a conoid which opens with Mach angle μ around the stream line,

$$\sin\mu = 1/M \qquad (52)$$

the Mach conoid, see Figure 13. Therefore all areas along Mach conoids from and towards each surface point have to be respected. In practical applications for slender configurations, the following double integral approximation is used to compute the area
distribution of an equivalent body of revolution: on the most important perturbation line
(e.g. the fuselage center line), the contributions of one Mach cone (of the undisturbed
flow) are collected by integrating the cross sectional areas which are cut by a sufficient
number of tangent planes to the Mach cone, so-called Mach planes (about 24 or more),
which osculate around the generating Mach cone; along the perturbation line (center line)
the contributions of several generating Mach cones (about 30 or more) are summed up
[51], see Figure 14. As long as the aircraft is slender, it does not strongly alter the speed
of the incoming flow; therefore the simple area approximation using the Mach planes of
the incoming flow is valid. An aircraft which produces strong perturbations, will never be
efficient. For simple bodies of revolution, the minimum drag area distribution is given by
the Sears-Haack area distribution [52], [53].

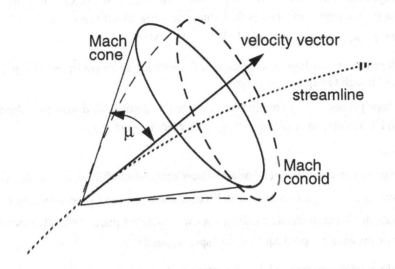

Figure 13   Mach Conoid and Mach Cone

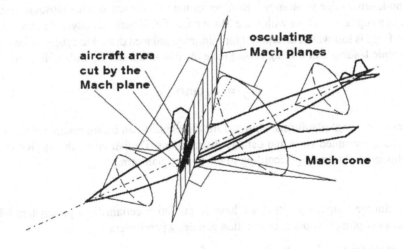

Figure 14   Wave Drag Calculation Using Mach Planes

Lift dependent wave drag:

The third term describes the lift dependent wave drag. It depends on speed (β^2 or β^2 / q), aircraft weight W and aircraft *lifting* length l.

Lift dependent wave drag rises quadratically with weight W^2 or C_L^2, like the induced drag in the last term. But whereas the induced drag lowers with speed (q), the lift dependent wave drag increases with speed ($\beta^2 / q \sim 1 - 1/M^2$).

Lift dependent wave drag lowers quadratically with lifting length l^2, requesting a slender wing with high sweep and small span.

The shape parameter K_L is minimized by a smooth elliptical lift distribution along lifting length l. Computation is demonstrated in other lectures and [54].

Induced drag:

The last term describes the well known induced drag. Its nondimensional value C_D, which is referenced to wing area S, rises with lift coefficient squared C_L^2 and decreases with aspect ratio b^2/S; but in physical units it rises with weight squared W^2 and decreases with dynamic pressure q (speed squared) and span squared b^2.

Therefore: minimum induced drag requires a high span (b^2).

The shape factor K_I is the well known Osswald factor, which becomes 1 for elliptical lift distribution over the span, see e.g. [55]. Elliptical lift distribution is achieved by plane wings of elliptical planform distribution and nearly by plane delta wings (as long as there is no leading edge separation). Best performance is reached with subsonic leading and trailing edges, i.e. edges within the Mach cone. For supersonic edges, in general the suction force is lost which is required for minimum induced drag. Pure supersonic wings (supersonic leading and trailing edges) have the maximum lift dependent drag of

$$C_D = C_L \tan \alpha \qquad (53)$$

For other (wing) planforms, elliptical lift distribution can be approximated for one angle of attack by suited twist and camber distributions. Deviations of this design α produce additional induced drag according to the (wing) planform.

Summary:

To minimize supersonic drag we have to minimize certain flow parameters which requires to optimize some configuration geometry parameters:

minimize *friction drag* by optimizing surface *geometry and quality* (laminarization),

minimize *volume wave drag* by optimizing *slenderness and volume*,

minimize *lift dependent wave drag* by optimizing *wing slenderness*,

minimize *induced drag* by optimizing *span, resp. aspect ratio*.

Wave drag minimization requires slender configurations and slender wings, i.e. low b/l, whereas induced drag requests for high span b. Task is, to find the best combination of b and l, represented by the size of the rectangle

$$b \cdot l \quad \text{or} \quad b \cdot l_0 \tag{54}$$

given by span b and lifting length l or overall length l_0 (see Figure 15).

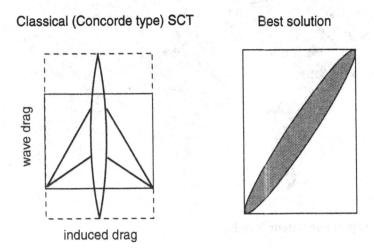

Figure 15   **Wave Drag and Induced Drag**

The classical Concorde-like configuration only provides relatively low span for acceptable slenderness with (at least partially) subsonic leading edges. Because the leading edge
is swept in different directions, it provides only half of the maximum possible lifting
length l for a given span b and sweep angle. In contrast, the oblique flying wing (OFW)
provides the maximum rectangle size $b \cdot l = b \cdot l_0$ because the wing uses the whole diagonal.
Indeed, the elliptical OFW is the best solution reaching the theoretical drag minimum
which was demonstrated by R.T. Jones [56].

4.3 Transonic Cruise

Because an SCT generates an annoying sonic boom as long as it flies faster than the speed of
sound, supersonic operation will be limited to uninhabited areas, i.e. mainly over sea. Over populated areas SCTs have to cruise at subsonic speeds. But many important airports like Chicago,
Atlanta, Denver, Frankfurt etc. are far away from sea and many routes contain an important part
of flight over populated land. Therefore routings over uninhabited areas are preferred, even at
slightly increasing distances; but large subsonic cruise legs remain, Figure 16. It is estimated, that

an SCT must be able to cruise subsonically during one fourth to one third of the whole distance.

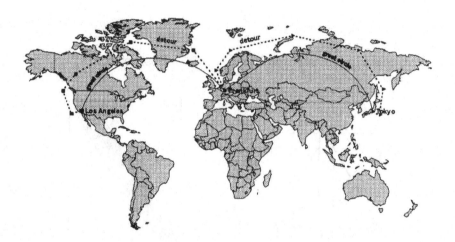

Figure 16   Supersonic Detour Routing

Therefore a new SCT must provide good transonic (i.e. high subsonic) cruise performance, in contrast to Concorde's poor subsonic performance. In addition to engines which provide good subsonic efficiency, aerodynamic design must provide this performance.

At high subsonic (transonic) cruise, good performance is achieved without significant contribution of wave drag, i.e. just below drag rise. So, in the drag composition of eqs. (49) and (50) the two wave drag terms can be omitted:

$$C_D = A/S \cdot C_f + K_I \cdot SC_L^2/(\pi b^2) \tag{55}$$

or in physical units

$$D = AC_f q + K_I \cdot W^2(\pi q b^2) \tag{56}$$

Here mainly the sum of friction and induced drag has to be minimized and to be balanced with the other design requirements.

The terms for friction and induced drag are described in chapter 2 after eqs. (49) and (50). Because here the flight is subsonic, supersonic leading edges do not exist. But optimum angle of attack is higher than at supersonic cruise. Therefore flow at relatively sharp or even really wedge-sharp leading edges may separate, producing high vortex drag which has to be avoided by suited leading edge flap deflections.

To minimize high subsonic drag, we have to

minimize drag components by to optimizing geometry components:

friction drag   **surface**, C_f (laminarization)
induced drag   **span** resp. aspect ratio, and avoid separation

Therefore subsonic airliners have highly loaded high aspect ratio wings, see Figure 17. This requirement contrasts to the supersonic cruise requirements. Here length 1 or l_0 is only required to delay the drag rise Mach number, and for flight stability and control. The best solution would be a variable geometry, which was proposed by Boeing for the first US-SST. But the high weight of large movable parts prevented such a solution. In contrast, the **Oblique Flying Wing** (OFW) [56] provides variable geometry without large moving parts; therefore it is the better solution.

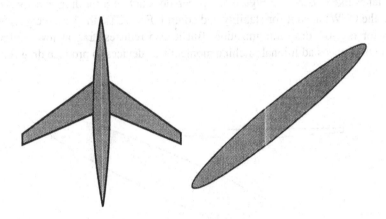

Figure 17   Transonic Cruise Aircraft

4.4 Take-Off and Landing

No aircraft can fly without take-off and landing. Therefore low-speed performance is crucial for the aircraft's viability. Low speed performance determines the required thrust during take-off and landing, and so strongly influences airport noise. For future SCT, take-off noise is the most difficult requirement! And a viable SCT must be even quieter than what is required today by certification authorities (stage 3), since many airports do not accept noisy climbing aircraft which comply to stage 3. They have to respect the concerns of the sourrounding communities and have to restrict noisy aircraft in order to maintain their airport certification.

At low speeds, high lift or high angles of attack are required. Wave drag does not exist, friction drag is nearly constant, so induced drag (and separation drag) dominates. This leaves only the last term in eqs. (49) or (50).

$$C_D = K_I \cdot S C_L^2 / (\pi b^2) \tag{57}$$

or in physical units

$$D = K_I \cdot W^2 / (\pi q b^2) \tag{58}$$

To minimize induced drag, we need a large span. This can be seen at high performance low speed planes like gliders, see Figure 18. Sweep does not help for drag, but special configurations (like the OFW) need it for stability and control. For SCTs, the variable geometry of the OFW helps for take-off drag minimization. But it also reduces drag at low landing speeds; therefore an OFW needs additional pitching moment free devices to produce drag for landing.

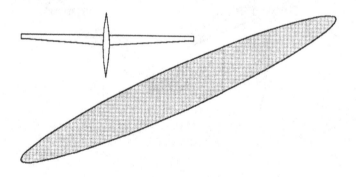

Figure 18   Low Speed Planes

Additionally we must avoid separation:

- Separation limits pressure recovery. This adds pressure drag. On unswept wings, nose separation produces sudden, unstable flow conditions which provide unstable flight conditions. This must be avoided for safety reasons. Controlled trailing edge separation -e.g. on flaps- can be used to increase drag, e.g. for a steper glide path or deceleration during approach.

- On highly swept wings, leading edge separation is controlled and produces additional vortex lift which is used to increase Concorde's lift. But these lifting leading edge vortices are positioned above the wing at strongly reduced lifting span compared to the wing span itself. This reduced lifting span increases drastically the induced drag, Figure 19. And Concorde needs its noisy afterburners to overcome this drag during take-off.

- A new SCT has to avoid leading edge separation during climb to enable low noise procedures.

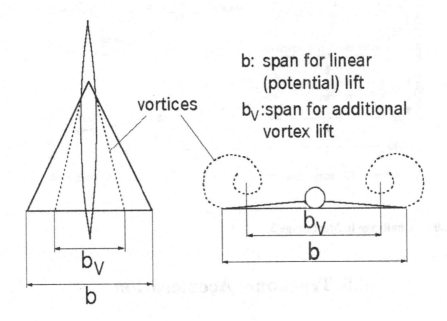

Figure 19   Separation Drag

Figure 20 shows a design exercise for a new "Concorde-like" SCT (for 250 passengers, 5 000 nm): Reference is a DA-design (DA: Daimler-Benz Aerospace Airbus) of an aircraft which is optimized just to meet stage 3. But because noise prediction is not very precise, the influence of aircraft noise on aircraft weight (MTOW) was investigated. Each square represents an optimized aircraft producing either more calculated noise than stage 3 (+ 0.5, to +9 dB) or less calculated noise (-0.5 dB). Weight strongly increases for lower noise, and just for -1 dB no solution was found. It has to be respected, that noise measurement at certification has a scatter of nearly 2 dB, calculated noise prediction about 3 dB and weight prediction has its scatter also. This means that our design is still more than marginal!

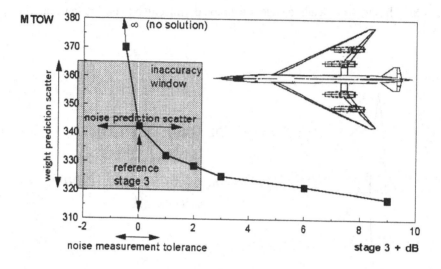

Figure 20   Challenge to Meet Stage 3

4.5 Transonic Acceleration

A supersonic jet is to cruise at supersonic speeds. But to reach supersonic speeds, it has to accelerate from take-off to the supersonic cruise speed. But in between there is a speed regime with maximum wave drag and minimum engine performance which may determine engine size. This critical speed is the transonic acceleration point at low supersonic speed (about Mach 1.1).

Because wave drag is at its maximum, it dominates drag and (for rough estimations) the other parts may be omitted. So eqs. (49) or (50) become:

$$C_D = K_v \cdot 128 V^2 / (\pi S I_o^4) + K_L \cdot \beta^2 S C_L^2 / (2\pi I^2) \tag{59}$$

or in physical units

$$D = K_v \cdot 128 V^2 / (\pi S I_o^4) \cdot q + K_L \cdot \beta^2 W^2 / (2\pi q I^2) \tag{60}$$

where the first term for volume wave drag is dominant.

At Mach 2 the Mach-angle is 30°, at Mach 1.2 about 60° and at Mach 1 just 90°. This means, that at Mach 1 all disturbances produced by the aircraft are radiated in a plane normal to the flight path and so stay in the relative position to the plane, at least when neglecting the local Mach number variations. (But these local Mach number variations are the reason, that stationary or very slowly accelerated flight at Mach 1 is possible).

To minimize transonic wave drag, the configuration must be slender with a smooth variation of the total aircraft's cross section area distribution. These cross sections must include the variations of the engine's stream tubes. For the large Mach angles of about 90°, changes of cross sectional area must be balanced in a very short streamwise distance; but due to the nearly stationary propagation, the area distribution within each cross section is not so important. This may lead to so called "Coke-bottle" fuselages which balance (strong) variations of wing, tailplane, engine areas by fuselage area changes.

To enable low wave drag at transonic *and* supersonic cruise speed, smooth area variations of all aircraft parts are recommended.

It is easy to design a perfect supersonic or hypersonic aircraft which will not be able to accelerate to supersonic speeds. This was demonstrated several times in the past! And Concorde needs its fuel guzzling afterburners to overcome this drag during a very slowly accelerated part of the flight. If Concorde does not succed in the first attempt to accelerate to supersonic speeds for a transatlantic flight, it has to return for refuelling!

4.6 Consequences

Contradicting requirements are daily life for an engineer. But for a conventional SCT these four requirements really pose a design trap, mainly:

- high supersonic cruise performance requires
    - a slender configuration with
  subsonic leading edges (i.e. limited span) or
  very thin wings with leading edge flaps or
  variable geometry
  low engine diameters,
- low take-off noise requires
    - large span with
    round leading edges or leading edge flaps and
    large engine diameters.

At DA a screening of several promising configurations was performed, see Figure 21:

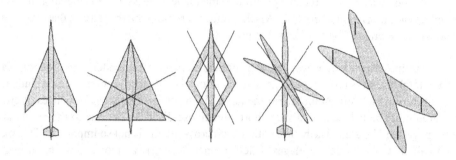

Figure 21   Search for a Viable Configuration

Remaining configurations were the conventional symmetric wing-body configuration at the left end, and the oblique flying wing (OFW) at the right end.

The blended wing-body lost for its poor slenderness. The optimizer simply concentrated the payload more and more at the center and spent more length for slenderness. So this configuration automatically transformed into the symmetrical wing-body.

The joined wing configuration has only limited span, does not provide enough fuel volume, does not provide space for an undercarriage and is structurally difficult, especially due to buckling.

The oblique wing-body combination is only interesting for rather small aircraft at low supersonic speeds [57].

The extremely different aerodynamic requirements seem to prohibit a solution for the conventional symmetric wing-body, at least for large aircraft and long range. Only limit performance of all disciplines' technologies may reach the limit of viability. (Besides aerodynamics, severe problems are e.g. weight, flutter, engines, long flexible fuselage, undercarriage). But this configuration has its merits for a smaller aircraft and shorter range, e.g. a 200 passenger transatlantic aircraft.

The OFW is limited to large passenger aircraft, because profile height must be about 2.5 m or more. Therefore it is suited for SCTs with more than 250 passengers and for subsonic aircraft with more than 400 passengers. The OFW provides variable geometry (aerodynamic span) without large moving parts, and best supersonic performance. It provides solutions to all the known problems of other configurations, like weight, noise, flutter, undercarriage, structural flexibility. In contrast it needs drag producing devices with controllable pitching moments to allow for a sufficiently steep descent and a short flare at touch down. Like with anything new there is still room for many new problems. Also, the interfaces between the individual disciplines are strongly different from conventional (subsonic) aircraft.

Both a conventional solution at design limits or an unconventional OFW-solution pose a strong challenge for aerodynamics and the other disciplines. The goal will only be met by new approaches using and further improving the techniques of Multidisciplinary Design Optimization (MDO).

Because both solutions require many new, unapproved technologies, flying technology demonstrators are required in preparation of civil passenger traffic.

4.7 References

[49] **Bruse, M., Bechert, D. W., van der Hoeven, J.G.Th., Hage, W., Hoppe, G.**
Experiments with conventional and with novel adjustable drag-reducing surfaces
Near Wall Turbulent Flows, R.M.C. So, C.G. Speziale and B.E. Launder (Editors), Elsevier Science Publishers B.V., 1993

[50] **FLIGHT INTERNATIONAL, 2-8 Nov. 1994, p. 24**
New Surface Coating Cuts Drag

[51] **Lomax, H.**
The Wave Drag of Arbitrary Configurations on Linearized Flow as Determined by Areas and Forces in Oblique Planes Ames Aeronautical Laboratory, NACA RM A55A18, 1955

[52] **Haack, W.**
Geschoßformen kleinsten Wellenwiderstandes
Lilienthal-Gesellschaft, Bericht 139, 1947

[53] **Sears, W. R.**
On Projectiles of Minimum Drag
Quart. Appl. Math. 4., No. 4, 1957

[54] **Van der Velden, A.**
Aerodynamic Design and Synthesis of the Oblique Flying Wing Supersonic Transport
PhD-thesis Stanford University, Dept. Aero Astro SUDDAR 621, Univ. Microfilms no. DA9234183, June 1992

[55] **Küchemann, D.**
The aerodynamic design of aircraft
Pergamon Press, Oxford, 1978/ 1985

[56] **Jones, R. T.**
The Minimum Drag of Thin Wings at Supersonic Speed According to Kogan's Theory
Theoret. Comput. Fluid Dynamics (1989) 1: 97-103

[57] **Van der Velden, A., von Reith, D.**
Multi-Disciplinary SCT Design at Deutsche Aerospace Airbus
Proceedings of the 7th European Aerospace Conference EAC'94 "The Supersonic Transport of Second Generation", Toulouse, 25-27 October 1994, paper 3.61

CHAPTER 5

REQUIRED AERODYNAMIC TECHNOLOGIES

J. Mertens

Daimler-Benz Aerospace Airbus GmbH, Bremen, Germany

5.1 Introduction

In the preceeding chapters on "Son of Concorde, a Technology Challenge" and "Aerodynamic Multipoint Design Challenge" it was explained, that a well balanced contribution of new technologies in all major disciplines is required for realisation of a new Supersonic Commercial Transport (SCT). One of these technologies - usually one of the most important for aircraft- is aerodynamics. Here, the required "pure" aerodynamic technologies are specified in more detail, according to our present knowledge. Increasing insight into the problems may change the balance of importance of the individual technologies and may require some more contributions. We must never confine our knowledge to the knowledge base of an expert at a given time, but must stay open for new insights.

5.2 Supersonic Flight Regime

5.2.1 Physics of supersonic flow

In air, information of small disturbances propagates at the speed of sound. This information is transported by collisions between the molecules due to molecular thermal motion building up a *bumping information chain* at the speed of sound. At subsonic speeds, this propagation speed is faster than flight speed. Therefore, the air molecules around an aircraft are informed about the motion of the aircraft by bumping neighbouring molecules. Changes of air velocity and pressure are therefore smooth at subsonic speeds.

At supersonic flight speeds, the aircraft is faster than the information within the air. If we look at the disturbances produced by one small disturbance moving at supersonic speed v_∞ in three succeeding seconds Figure 22, the disturbed field is inside of a cone, the Mach cone. Outside this cone, air is not yet informed about the disturbance and nobody can hear anything of it. Important parameters are:

$$M = v/a, \qquad M{:}Mach\ number \tag{61}$$

$$\sin\alpha = a/v = 1/M, \qquad \alpha{:}Mach\ angle \tag{62}$$

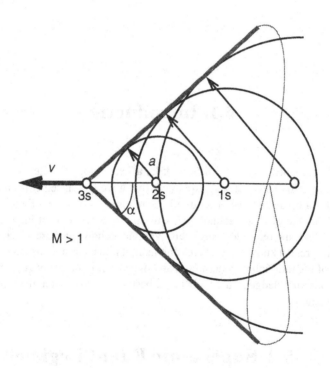

Figure 22   Mach Cone

Information is spread only within the Mach cone. But the transport of individual informations from neighbouring point to point is even more confined, Figure 23:

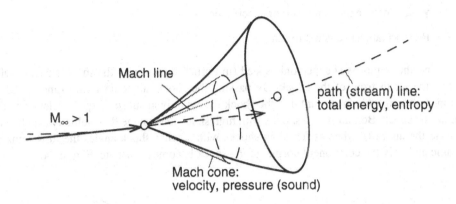

Mach line

$M_\infty > 1$

path (stream) line:
total energy, entropy

Mach cone:
velocity, pressure (sound)

Figure 23   Propagation of Supersonic Perturbations

• Information about small perturbations in pressure and velocity is transferred only along Mach-lines in the Mach cone.

• Information about small perturbations in total energy and entropy is transferred only along the path line or stream line (stationary flow).

The formulations above reflect at first only inviscid flow without thermal conductivity. Viscous fluxes and heat conduction exist normal to the Mach cone and path (stream) line. These additional fluxes become important along the path (stream) line; along the Mach cone their influence is very low.

An aircraft flying at supersonic speed is not a small, but rather a strong perturbation. Because the air in front of the aircraft cannot be informed about the coming perturbation, smooth reaction is impossible. Instead, the air reacts instantaneously on this strong perturbation and adapts with a strong reaction to this situation: Within some free path lengths of the molecules (about 5 to 7) the airflow changes its mean velocity direction to follow the aircraft's surface. This sudden reaction not only requires a rapid flow change at the aircraft's surface, but also is radiated into the flow field, Figure 24.

Air can transport information in smooth regions of the flow field only at the speed of sound, i.e. inside the Mach cone. This requires that

• the Mach angle in front of the shock is smaller than the shock angle (otherwise a shock would not be needed, because smooth information would be possible);

• the Mach angle behind the shock is larger than the shock angle, because otherwise no information would be available to build up a shock.

Speed of sound, a, only depends on temperature T:

$$a^2 = \gamma R T \tag{63}$$

γ: adiabatic exponent (ratio of specific heats)

R : special gas constant for air

The shock moves at supersonic speed (in normal direction to the shock surface) with respect to the air in front of the shock. By passing the shock, air temperature (and speed of sound) increases to such an amount that the shock moves only at subsonic speed relative to the air behind the shock. Behind strong shocks with high shock angles, the flow velocity is subsonic (relative to the aircraft), whereas for weak shocks with smaller shock angles the flow remains supersonic and only the component normal to the shock becomes subsonic, Figure 24.

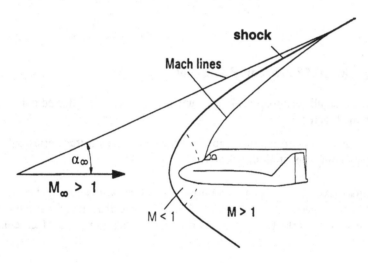

Figure 24   Strong Perturbations

Shock energy remains in the shock surface and is radiated only along the shock surface. Furthermore, by conflicting information from the Mach cones in front and aft the shock, new energy is radiated into the shock. Shock strength in larger distance from the aircraft therefore only very slowly decays.

An SCT flying at about 16 000 m altitude produces a strong shock on the ground, the sonic boom Figure 25. Usually, the pressure history of the sonic boom shows two shocks: a front shock, a rear shock, and in between a nearly linear decrease of pressure, the so called N-wave. All information of zones with higher temperature concentrate at the front shock; all other informations of the regions with decreasing temperatures are collected by the rear shock. So the stable N-wave builds up and can be heard as a double bang on the ground. It is possible to design pressure distributions around the aircraft which do not steepen up to the pure N-wave, but those pressure distributions are not stable. Only minor changes in temperature distribution of the air (weather conditions varying stubstantially in real atmosphere) or flight conditions

(Mach number; lift coefficient) destroy any carefully tuned pressure distribution and an N-wave becomes dominant.

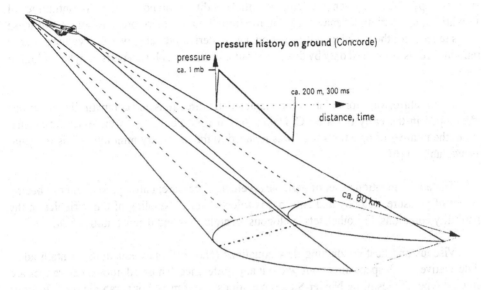

Figure 25   Sonic Boom

The sonic boom is always to be heard when an aircraft passes flying faster than speed of sound. The bang is the stronger the heavier and shorter the aircraft is, and the lower it flies. The boom carpet has a limited lateral dimension of about 80 km (as the Concorde), because higher speed of sound (temperature) at low altitudes produces extinction of disturbances. This lateral carpet size strongly depends on weather (temperature distribution) and speed of the aircraft with respect to the ground. Boom strength usually is strongest about the middle of the carpet; at the side of the boom carpet, noise is lower and softer. Outside the cut-off distance no bang can be heard, but -if any- only the usual aircraft noise, like a grumble.

5.2.2 Mathematics for solving supersonic flow problems

Mach cone and path (stream) line, sketched in Figure 23 are singular surfaces called "characteristic surfaces"; their generating lines are Mach lines or "characteristics". Inviscid supersonic (or transient) flow is completely described by a set of partial differential equations (PDEs), valid only along characteristics, called "compatibility equations"; they do not contain any derivatives across the characteristic surfaces, but allow for undefined jumps in these derivatives [58], [59], [60], [61], [62]. This means:

Any other set of PDEs describing supersonic (or transient) flow contains *derivative components* normal to characteristic surfaces which are *not defined by the PDEs*! Solutions of those equations may use invalid information or produce solutions containing random parts. This may prohibit accurate or even useful solutions.

Inviscid supersonic (or transient) flow equations are hyperbolic. They describe radiation problems. The PDEs itself allow for discontinuities in the derivatives of the variables (like velocity, pressure, entropy). If, for a given problem, the initial conditions do not contain discontinuities of the derivatives, discontinuities may evolve in the flow field. Furthermore, any solution to these equations (except for the trivial identity solution, i.e. all derivatives are given everywhere as zero and remain zero) is composed only by discontinuous elementary solution parts, maybe for higher derivatives.

A straightforward formulation for characteristic directions and compatibility equations was developed in the early 50ies by C. Heinz [63] at ISL: Focusing on the essential normal direction, the number of equations used was reduced to the necessary minimum. This formulation is available in [61].

The above mentioned set of variables (velocity, pressure, entropy; total energy beeing dependent of pressure, velocity and entropy) is selected for decoupling of the variables in the compatibility equations. For other sets analogous formulations and discontinuities hold.

Viscous and heat conducting flow equations (Navier-Stokes equations) contain additional derivatives in all space directions without any preference. Those additional derivatives are of elliptical type; the resulting Navier-Stokes equations are of mixed or parabolic type. Viscous and heat flux influence is limited to thin layers (boundary layer, shear layers, shocks) and separation regions.

Shocks can develop in the flowfield by steeping up of solutions and at boundaries, where sudden changes of boundary conditions occur. Shocks are described by the Rankine-Hugoniot equations [58], [59], [61] which are derived by surface integration over a flat volume along the shock surface. Even the "inviscid" Rankine-Hugoniot equations contain viscous and heat fluxes across the shock surface. So the whole flowfield can be described by the inviscid equations, except boundary layer, shear layers and separation zones. In the free flow field, the discontinuous solution properties of the "inviscid" equations must be respected. Even when solving the Navier-Stokes equations for shocks, the thickness of a shock (less than 10 molecule free path length) is below numerical resolution; within this small layer the number of molecules is not sufficient to establish state variables as required for the continuum formulation of the Navier-Stokes equations. Therefore validity of solutions must be carefully checked.

In frequent case studies the capturing of shocks in numerical solutions is improved by selection of so called conservative variables which should be conserved when passing a shock. Caution is needed, though: In the Rankine-Hugoniot equations, basically not the variables are conserved, but their fluxes normal to the shock. For example, normal to a stationary shock not density ρ is conserved, but $\rho v n_s$, with n_s the shock normal vector; only by chance, the conservative velocity ρv (i.e. momentum) is the flux of ρ. On the other hand, v_t, the velocity component parallel to the shock surface, is conserved across the shock, but not ρv_t.

5.2.3 Dominating flow phenomena for SCTs

Wave drag:

In supersonic flow the disturbances are radiated away from the aircraft surface. Pressure balancing between aft and forward flow is impossible (Figure 26) or strongly limited for winged vehicles. The result is wave drag, corresponding to the radiated energy [64].

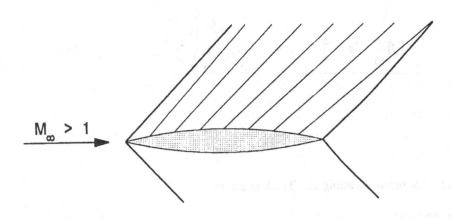

Figure 26   Wave Drag

All energy is radiated along Mach lines. Pressure (and temperature) changes along the aircraft, resulting in a crossing over of Mach lines at some distance. When Mach lines intersect, conflicting information arrives at this point which will be bridged by a shock. Eventually all wave drag energy is captured by shock energy.

Circulation:

Disturbances can only propagate within the downstream Mach cone. This limits a build-up of circulation for finite wings (Figure 27). Leading edge flow can only influence the downstream part of the leading edge if the leading edge stays within the Mach cone (this is a so called *subsonic leading edge*). Information of the trailing edge can only reach other parts of the trailing edge to improve pressure recovery if the trailing edge is located within the Mach cone (*subsonic trailing edge*). The trailing edge can only improve a build-up of circulation if parts of the leading edge are within the trailing edge's Mach cone; this is only possible for low supersonic Mach numbers, high aspect ratio and high sweep angles.

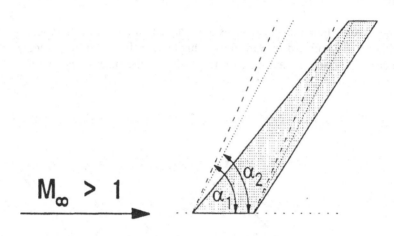

Figure 27   Subsonic Leading and Trailing Edges

Kogan's theorem:

Lift efficiency in subsonic flow is described by the downwind field behind the aircraft: size, strength and downwind distribution determine the induced drag [64]. Kogan [65], [66] developed a similar theorem for supersonic flow: Any point of the aircraft surface can only influence air in its downwind Mach conoid; any point on the aircraft surface can only be influenced by air in its upwind Mach conoid. Kogan constructs the envelope of all downwind Mach conoids originating at the aircraft leading edges, and all upwind Mach conoids originating at the aircraft trailing edges. The control surface defined by the intersection of those two envelopes contains all downwind information of the aircraft. Lift-dependent drag of an aircraft is the smaller the larger this control surface area is, the smaller the mean downwind is and the less disturbed the downwind distribution is.

Interference drag:

Shocks generated on the surface of engine nacelles (and other interfering parts) are radiated to neighbouring parts like the wing, to other nacelles or the fuselage, where they are reflected (Figure 28). Reflection conditions and drag is strongly influenced by shock-boundary layer interference. This requires reliable nonlinear calculation methods including viscous effects. Wind tunnel simulation must be able to simulate the viscous effects of high flight Reynolds numbers.

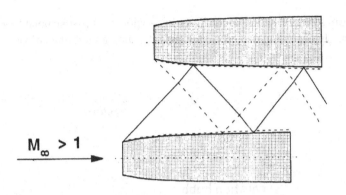

Figure 28   Shock Reflection on Neighbouring Engines

Inlet flow:

Jet engines work at purely subsonic speeds. The inlet of an SCT, therefore, has to decelerate incoming air from supersonic to subsonic velocities. This requires passing a shock system. To minimize shock losses, the air passes through several shocks (possibly including some isentropic compression). Adequate mathematical models for the generation of shocks, control of shocks, shock position and shock reflection, including important viscous effects are strongly nonlinear. To enable stable flow conditions and engine operation, the inlet flow must be balanced with the nozzle flow; Concorde's aerodynamically coupled nozzle and inlet control is still state of the art (Figure 29). Highly sophisticated numerical and wind tunnel simulations are required and must be combined with the control system and engine operation.

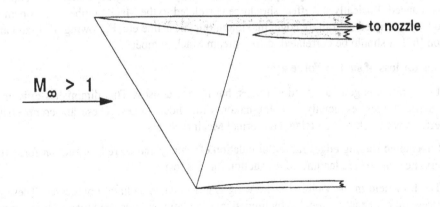

Figure 29   Inlet Flow, Concorde principle

Hinge line shocks:

At control surfaces, shocks at the hinge line region can provoke boundary layer separation bubbles strongly degrading flap efficiency and introducing vibrating airloads (Figure 30).

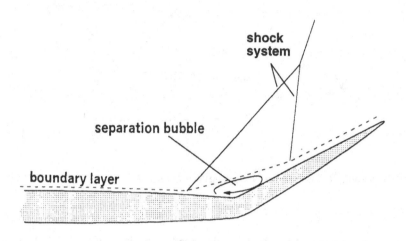

Figure 30   Shock-Boundary Layer Interference

5.2.4 Unresolved problems

Suction force:

Compared to lower flight speeds, the efficiency of leading edge suction in supersonic flow is reduced. Basically the effect should be correlated to the Mach number components normal to the leading edge, trailing edge, maximum thickness line etc. Following M. Mann and H. Carlson [67], it should be correlated to free stream Mach number.

Reasons for loss of suction force are:

- Low pressures generate suction forces, but reduce density. This diminishes efficiency of suction forces, especially in combination with shock losses. These are compressibility effects which should be related to normal Mach numbers.

- Supersonic trailing edges inhibit circulation efficiency and so reduce suction force recovery. For supersonic leading edges suction force is lost.

- The flow field in the vicinity of the wing is governed by radiation processes. These processes are not axactly modeled by numerical calculations: linearized theory does not model effects of local Mach number variations and therefore is unable to produce the correct radiation directions (characteristics). Most nonlinear methods respect for the local Mach numbers, but do not exactly model radiation; so numerical diffusion smears out radiation transport.

- Supersonic wings are mainly designed for minimum wave drag. This leads to nearly conical flow situations. At higher Mach numbers with smaller Mach angles this introduces strong pressure gradients in spanwise direction, i.e. normal to the free stream direction. Boundary layer flow tends to follow local pressure gradients; so, boundary layer air will accumulate in the low pressure valleys on the wing and may modify the designed low wave drag pressure distributions. This effect should strongly depend on Reynolds number and might be stronger in low Reynolds wind tunnel tests than in free flight. This effect is mainly related to free stream Mach number.

Radiation in CFD solutions:

Linearized theory does not model effects from local Mach number variations. Usual CFD methods do respect these, but only marginally model radiation properties. Numerical stability is achieved usually by addition of numerical viscosity. Without proper modelling of radiation properties, though, random contributions are introduced into the solution or valid contributions are assumed to be zero. Upwind schemes should model radiation, but most upwind schemes are basically one-dimensional and cannot model radiation direction, like all the upwind schemes which only fulfill the eigenvalue sign; i.e. they approximate the radiation direction by an accuracy of up to $\pm 90°$. Only CFD methods which are carefully based on the method of characteristics provide good radiation properties, but these methods usually are not suited for universal CFD codes, especially for their rather inflexible handling of complicated geometries. A challenge remains to improve the tools for supersonic CFD.

Physical drag contributions:

To improve the aerodynamic design of an aircraft, it is very helpfull to know the different contributors fo physical drag:

- Wave drag (radiated energy plus entropy generated by shocks)
- induced drag
- friction drag
- separation drag.

For subsonic flow and linearized supersonic flow, methods exist largely based on far field balances, e.g. [68], but for nonlinear supersonic flow the far field results are poor because of inexact radiation models. Also, surface integration accuracy is more difficult to achieve; only friction drag can easily be extracted.

Supersonic laminar flow:

This technology is still a big challenge, for both theoretical predictions and even more for experimental verification. With aerodynamic efficiency improvements by successful laminarisation promising to be very high, a special book chapter is devoted to that field.

5.2.5 Tasks

The following tasks summarize the challenges of supersonic aerodynamic design:

- Provide aerodynamic data suited for interdisciplinary design optimisation. These data need not result from best achievable accuracy limits, but must be reliable within specified accuracy conditions for a wide range of configurations.

- Maximize aerodynamic performance (Lift/Drag = L/D) for given geometrical constraints: improve quality of aerodynamic tools to better reflect flow physics, balance wave drag, induced drag, friction drag (including laminarisation concepts) for minimum overall drag.

- Determine the limits of special flow phenomena, like suction force, etc.

5.3 Low Speed Flight Regime

5.3.1 Dominating flow phenomena

At take-off and landing, low speed of the aircraft generates only small dynamic pressures. Generation of aerodynamic forces, therefore, requires high specific aerodynamic loading, up to the separation limits.

Low speed lift generation:

To improve lift, span can be increased (variable geometry aircraft like the OFW), camber can be increased by flaps and wing area by Fowler flaps. Angle of attack can be increased (especially for highly swept wings) and engine air can be used directly for lift generation or for support of flap efficiency. Most SCT configurations have only limited possibilities to use flaps: the symmetric (Concorde-like) configurations have large wing areas, where flaps can contribute only marginally. For an arrow wing, inner wing flaps can be efficient, because they generate lift near to the center of gravity which reduces trim losses. For the OFW the pitching moments connected with camber or fowler flaps limit their application.

Of special importance are leading edge flaps for symmetric SCT configurations which do not generate lift but reduce lift dependent drag. Concorde generates additional lift using the lifting vortices generated by leading edge separation on highly swept wings. Those vortices, on the other hand, produce high drag. For a new SCT it is intended to use those vortices - if used at all - only at lift-off and perhaps flare. During climb, no vortices should separate at the leading edges to allow for lower climb drag. There are proposals to use droop flaps (Figure 31), or - proposed by Boeing - suction at the leading edge to delay leading edge vortex separation.

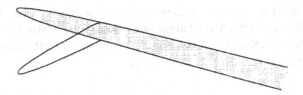

Figure 31   Drooped Leading Edge

Separations:

At high aerodynamic loadings separation may occur. For landing, separation with separation drag is welcome, but separation must always be controlled; it must not suddenly alter the flight handling. Leading edge separation must therefore be confined to highly swept leading edges, where the individual separation vortices are fit to the large lifting vortex. Trailing edge separation has to occur smoothly and at selected parts, like for subsonic aircraft. Especially the OFW needs much drag for landing to inactivate its superior aerodynamics. Drag producing devices are then requested which introduce only minor pitching moments.

Controllability:

Due to the small dynamic pressure, control surfaces become less efficient. In addition, separation on control surfaces limits the achievable forces; separation on wing, fuselage and nacelles introduces additional disturbances.

Especially for highly swept trailing edges - like for arrow wings and OFW- the effectivity of control surfaces is not yet completely undersood. Some additional research is needed here.

In order to exploit the wing's lift performance with droop leading edges and cambered or even fowlered trailing edges, it is necessary to balance the aircraft by an additional control surface like a horizontal tail or canard.

Performance:

At take-off, main emphasis is on good L/D to reduce thrust and noise. For landing, though, high drag is necessary to allow step descent and slowdown, when the engines still run at flight idle with not too low thrust levels. On the other hand, enabling flare or allowing for go-around, drag must stay below some limits or must rapidly be reduced.

5.3.2 Mathematics

(Steady) subsonic flow problems are mainly of elliptic type in both the inviscid and viscous parts, which means that functions and all derivatives are continuous. Only the inviscid Euler equations allow for discontinuities normal to the path (stream) line. Methods are on the development level as available for subsonic aircraft. Similar to subsonic aircraft technology, extremely complex flow separation is not yet really understood.

5.3.3 Tasks

At take-off, it is necessary to

- improve lift

- maintain control

- improve L/D by reduction of drag;

whereas at landing it is required to

- improve lift

- increase drag, possibly by drag control devices

- guarantee handling qualities, especially when using partially separated flow.

5.4 Transonic Flight Regime

5.4.1 Main properties of physics of transonic flow

If the flow field contains parts with subsonic flow and other parts with supersonic flow, the problem is called *transonic*. Properties in the flow field strongly change:

- In the subsonic parts, the flow field balances all flow properties in the field and is described by elliptic *balance* equations.

- In the supersonic parts, the flow field cannot balance properties because flow speed is higher than information speed; it is described by hyperbolic *radiation* equations.

The supersonic parts usually are terminated by a strong shock. Whereas supersonic oblique shocks usually are kinematically fixed to the geometry, the strong normal shocks in transonic flow evolve in the flow field and are controlled by flow forces. Therefore they are prone to oscillations.

5.4.2 Mathematics of transonic flow

Mathematical models describing gas dynamics of those different flow types are strongly nonlinear, the equations are of mixed elliptic - hyperbolic type, even for the inviscid parts. Computational analysis usually follows transient formulations to model type changes within the flow. These enable elliptic, hyperbolic and parabolic solution procedures. Some models may be used for design purposes to calculate flows with desirable aerodynamic properties, see book chapter 7.

5.4.3 High subsonic cruise flight

Supersonic aircraft is firstly optimized for supersonic cruise performance. To achieve good transonic (i.e. high subsonic) cruise performance, it can be necessary to adapt the configuration to the different requirements. The OFW performs this by adaption of sweep. A symmetric configuration (like Concorde) can use flaps:

• leading edge flaps, to avoid leading edge separation; this is especially important for sharp (supersonic) leading edges,

• trailing edge flaps to control lift distribution for minimization of induced drag and load control during manoeuvers; this requires a pitch control surface like a horizontal tail.

5.4.4 Low supersonic acceleration flight

During transonic (i.e. low supersonic) acceleration, wave drag is dominant. Here it strongly depends on interference of the different parts of the aircraft: fuselage, wing, nacelles with engine stream tubes, tail. Near Mach 1 even small changes in Mach number produce strong changes in Mach angle (i.e. radiation direction); strong shocks prevail with considerable shock-boundary layer interaction. As a consequence, accurate drag prediction of the transonic interference phenomena requires nonlinear methods including the simulation of viscous effects.

5.4.5 Transonic control

Around flight Mach number 1 strong normal shocks are generated. These are very sensitive to small changes of the flow field and tend to oscillations.

Control flaps have to produce aerodynamic forces by pressure differences between the flap sides and adjacent wing area. Usually they generate a pressure rise on one side and a pressure drop on the other. Near Mach 1 the pressure rise at the hinge line provokes a strong shock with strong boundary layer interference. This easily results in vibrating loads (buffet) and weakens the flap's control forces. During transonic acceleration, therefore, the aircraft should not require strong control forces. Suited control flaps have swept hinge lines, hinge lines in less critical regions (for instance close to the trailing edge of the wing), or moving tails.

Engine efficiency is critical at low supersonic speeds. Especially the inlets have to cope with rapidly varying conditions due to sensitive Mach angle variations. The shock system, designed for supersonic cruise, cannot yet establish; an inlet control mechanism, designed only for supersonic cruise shocks, does not work. Measures are required, therefore, to allow for sufficient inlet efficiency, like special inlet doors.

5.5 Tasks:

At high subsonic cruise optimize L/D:

- avoid separation,
- minimize induced drag.

For transonic acceleration:

- minimize wave drag which is dominated by interference effects,
- provide control of the aircraft,
- provide control of engine inlet and nozzle.

5.6 Flap Effectiveness

5.6.1 Supersonic hinge lines

At control surfaces with supersonic hinge lines, shocks occur at the hinge line. Shocks produce pressure losses and so reduce flap effectiveness. Additionally, the shock can provoke boundary layer separation bubbles (Figure 30); pressure in those bubbles is lower than behind the final shock. This reduces the flap force significantly. Because of the system of the three shocks behaving very sensitive to variations in the incoming flow and to fluctuations in the separation bubble, strong vibration loads can arise.

5.6.2 Trailing edge flaps on highly swept wings

On wings with highly swept (subsonic) trailing edges (OFW, arrow wings), the boundary layer is deflected by the spanwise pressure gradients and tends to become nearly parallel to the trailing edge, or even separates (Figure 32). Tendencies known from lower sweep angles, and results for very high sweep angles, are not conclusive. Further theoretical and experimental investigations are required to understand flap efficiency at relevant sweep angles.

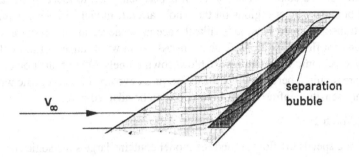

Figure 32   Highly Swept Trailing Edge

5.7 Wind Tunnel Measurements

Wind tunnel experiments are essential

- to get insight into still unknown flow physics, like separation, turbulence, transition,

- to validate numerical calculations,

- to generate data for complicated configurations including interference effects,

- to check aerodynamic design cases and to generate data for improvement strategies,

- to establish aerodynamic data for pre-flight validation of new aircraft.

Wind tunnels have limitations as well. For SCT development the most important limitations for wind tunnel investigations are:

Reynolds number:

Reynolds number in wind tunnel testing usually is an order of magnitude lower than in free flight, for supersonic testing often up to two orders of magnitude. For drag measurements the boundary layer is tripped; i.e. transition strips provoke transition from laminar to turbulent flow at defined positions. This allows for calibrated friction measurements, but the boundary layer is thicker at lower Reynolds numbers. Therefore the interference effects, especially shock/boundary layer interference, in the wind tunnel remain different compared to free flight. Technologies to transpose interference prediction from wind tunnel to free flight need to be developed.

Transition control:

Transition strips must be as small as possible. Thick or wide transition strips generate too much strip drag and thicken the boundary layer. On the other hand, if transition strips are too small, no transition or even relaminarisation occurs which does not allow useful drag measurements. The control of transition in experiments, i.e. to identify the laminar and turbulent boundary layer regions on the whole aircraft model is always required. Most common transition control methods - like the acenaphtene technique - require wind tunnel runs at constant flow conditions. This is impossible in wind tunnels of blow-down type; and many supersonic wind tunnels are blow-down tunnels. Other transition control technologies are required here; possibly the techniques devoleped for cryogenic wind tunnels (like highly sensitive infrared measurements) can be adapted here.

Testing around Mach 1:

At near sonic speeds the flow around the model contains large supersonic flow domains. In the supersonic regions the wind tunnel model radiates disturbances to the wind tunnel wall and are (at least partially) reflected by the wall back onto the model. In contrast to free flight conditions, this reflection strongly changes pressures and flow properties at the model.

At high subsonic speeds, the supersonic regions can reach the wall and so generate a choked supersonic nozzle flow over the aircraft instead of the open supersonic bubble over the free flying aircraft. This (partial) nozzle flow changes the whole flow field and does not further resemble to free flight conditions.

Most transonic wind tunnels have slotted or perforated walls in order to minimize wall reflections. This minimisation, though, is only sufficient if the supersonic bubble does not reach the wall or the important reflections do not meet the model. This requires test flow conditions avoiding the vicinity of Mach 1. New transonic wind tunnels use flexible (adaptive) walls, where the wall geometry is adapted during the test to follow a free stream path line. This allows for better adaption of near sonic test conditions, but quality of adaption depends on the technical concept of the adaption mechanism; usually only a plane wall adaption is possible for two of the four walls surrounding the test chamber. Although two-dimensional adaption in the most important direction is much better than no adaption, three-dimensional adaption for three-dimensional models remains impossible.

Engine simulation:

Usually, engines in supersonic tests are modelled by simple through-flow nozzles, but it seems impossible to design spillage-free through-flow nozzles: nozzles are choked, whereas an engine adds energy. Wind tunnel simulation is restricted therefore to cases including spillage. Additional nozzle base drag is created due to the choked flow. It can be corrected by pressure measurements on the nozzle, but these will correct only the individual nozzle base drag, not the additional interference wave drag.

Laminar tests:

There is not yet any wind tunnel available for supersonic laminar flow tests between Mach 1.5 and 2.5. At the University of Stuttgart, Germany, a facility is refurbished which - hopefully - will be suited. See also the special chapter 18 on laminar flow for supersonic transports.

Measurement techniques have been developed for exploiting wind tunnel experiments. Some of these techniques are state of the art and provided by all wind tunnels: force measurement and pressure measurement via small holes in the model surface. The simpler optical methods like shadowgraphs, Schlieren or interferograms are best suited for 2D-measurements and available where suited. More refined techniques are available and will be applied to supersonic testing, especially optical methods for flow field measurements:

Pressure sensitive paint (PSP):

Special paints are developed which, when illuminated by a special light source, emit light depending on the amount of O_2-molecules embedded in the paint surface. In air the amount of O_2-molecules directly correlates to air density. This allows for a direct measurement of air density distribution on the model surface and, when temperature is known, indirectly for the measurement of pressure distribution, see. e.g. [69], [70], [71]. This technique is new and needs further improvements before it can be applied as a stand alone pressure measurement technique. Especially paint thickness or durability, painting, illumination technique and related automatic data processing need further research.

Liquid crystal coatings:

Surface coatings based on liquid cristal technology allow for various mapping techniques of relevant flow parameters on the model's surface like shear stress and temperature. This allows simultaneous measurement on large parts of the model. Problems result from the relatively rough coating surface, the limited view angles and often the multiple sensitivities of the coatings which require careful separation of the measured effects. These problems presently still allow only for limited use of the technique in aerodynamic measurements.

Distribution measurements in the free flow field:

Several new techniques allow for measurements in a selected plane of the flow field. Most common is Particle Image Velocity (PIV), see e.g. [72]: The flow field of interest is seeded with microscopic particles, commonly droplets of about 1 µm diameter. In the plane of interest, those droplets are photographed twice within a short time interval. The movement of the droplets is identified to provide the droplet's speed which is equal to flow velocity except within a shock. New developments are aimed at larger measurement fields and measuring all three velocity components. In the future, PIV measurements will be usable even for complicated interference flow measurements.

Several other flow field measurement techniques are under development but either still in their infancy, restricted to high Mach numbers, suitable only for very specific cases or just of poor accuracy.

Laser-Doppler anemometry:

In the last years Laser-Doppler Anemometry (LDA) was developed as a tool for accurate, pointwise, nonintrusive flow field measurements [73]. The flow is seeded with droplets (like for PIV) which are observed within a small measurement volume. This volume is established by the crossing of two laser beams, where interference produces a sequence of light and shadow like a grid. Observed is the motion of the droplets through the interference grid, where the frequency of reflected light spots is correlated to the droplet's speed. LDA allows for accurate measuring of mean and fluctuating values, even resolution of boundary layer flow, but it requires relatively long measurement time.

Complementary to the experimental measurements correction methods for experimental errors or insufficient simulation are needed. The most important corrections required are:

- accurate correction of wall interference,
- correction of Reynolds effects, especially for interferencies,
- spillage and nozzle base drag correction, especially for flows without spillage.

5.8 Aeroelastics

5.8.1 Static aeroelastics

In classical aerodynamic design, the aircraft shape is designed for one design point M_C (cruise Mach at a given weight and altitude). Knowledge based margins provide the ability to cope with the off-design points. Some still cover the (cruise) flight regime (like M_{MO} = maximum operating Mach, M_{CS} = subsonic cruise Mach, other aircraft weights for begin of cruise or end of cruise, altitude variations ...). Others concern exceptional points which do not occur in normal cruise, but only e.g. for emergencies like M_D (dive Mach).

Aerodynamics assume the geometry to be rigid. Once the aerodynamic shape is fixed and the aerodynamic loads are known, structure loads are determined, structure is designed and static aeroelastic deformation is calculated. This deformation at the design point is taken into account when the shape to be built is defined. The procedure reestablishes the designed aerodynamic shape at the design point flight loads (M_C, design weight and altitude). For any deviation of the design point the aircraft will have a different shape. This deviation becomes important if the wing is not very stiff and if the deviations from the design point are large. Both occur for SCTs with thin wings and multipoint design conditions.

To find the best compromise for an elastic wing flying at different design points, aeroelastic deformation must be considered in the aerodynamic design. For aerodynamics this can be a rather simple formulation, like a beam formulation for a slender arrow wing or an OFW, or a simple shell formulation for some kind of delta wing, including bending and torsion. The difficult problem is the "simple" estimation of structural values because this requires simultaneous estimation of loads, mass distribution and structural thicknesses.

5.8.2 Flutter

Concorde has inacceptable take-off noise levels. For a Concorde-type SCT lower noise levels can only be achieved using larger engine diameters and larger wing span. To maintain or even increase cruise performance, wing thickness will have to be reduced. Such wings become very flutter sensitive.

Aerodynamic damping is an indicator for flutter onset. It is the smaller, the higher the flight speed is. At high subsonic speeds, nonlinear transonic aerodynamics reduce aerodynamic damping, the so called transonic dip (Figure 33) [74]. A new SCT has therefore to be investigated for flutter at transonic and supersonic cruise speeds.

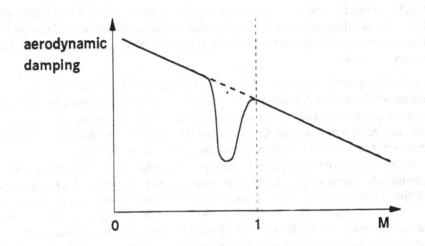

Figure 33   Transonic Dip

Because of flutter becoming very critical for the thin wings of symmetric (Concorde-like) SCT-configurations, at least a rough approximation of flutter tendencies must be included in the first steps of configuration optimisation. Hitherto nobody knows how to do it. Perhaps artificial flutter damping can help, if its certification becomes possible.

5.9 Geometry Generation

Aerodynamic design is development of a suited shape. For SCT development, extensive application of numerical optimizers is required. When using optimizers, the first very important step is to describe the space of possible shapes by as few parameters as possible, but still without in-*acceptable restrictions*. In the first step of interdisciplinary optimisation, only global parameters are needed to described the basic aircraft geometry. The more refined the investigations are, the more sophisticated the numerical methods are, the more detailed the geometry must be described. But for all levels the same requirements for geometry generation hold:

- For geometry generation by a human design engineer:
  Geometry must be described by a limited set of parameters. But those parameters must be meaningful and well ordered in order to allow a human to reach geometric design goals. Alternative ways are allowed, e.g. multiple parameter sets or parameter set selections.

- For automatic geometry generation by a numerical optimizer:
  Geometry must be described by as few independent parameters as possible. Those parameters may have any level of abstraction. Not allowed are alternatives to the optimizer for selection between different, but equivalent parameter sets.

- Any geometry generator must provide smooth shapes without tending to wiggles:
  If wiggles cannot be avoided, smoothing procedures must be provided. For human applications, the smoothing procedures can be applied off-line as the last step of geometry generation. For numerical optimizers smoothing, if not avoidable, must be included in the geometry generation.

- Any geometry generator must provide interfaces to and from CAD-systems:
  When aerodynamics has developed a shape, this shape will be transferred to other company work groups like project, structure, aeroelastics, model design and fabrication. All aircraft related data transfer uses CAD-systems. The aerodynamic shape therefore has to be transferred into the CAD-system without intolerable accuracy losses.
  On the other side, aerodynamics has to use input from other departments for geometry constraints like fairing size etc. Or the real model geometry has to be checked prior to a wind tunnel test. Or geometries generated by a partner must be investigated. Or wind tunnel results - like pressure measurements - have to be applied to a given geometry for aerodynamic improvements. In all those cases it must be possible to transfer the CAD-geometry into the aerodynamic geometry generator as an input geometry, e.g. to start an improvement calculation.

Especially for application of numerical optimisation strategies more progress in systematic shape definition is needed. Sometimes it is proposed to use CAD-systems directly for geometry generation, but CAD-systems are oriented towards structural design: these do not contain geometry definition tools suitable for aerodynamic optimisation. Powerful aerodynamic 3D-geometry generators are under development as preprocessors for CAD systems, see book chapter 9.

5.10 Fast Computer Codes for Aircraft Design

In the first interdisciplinary design loop, the whole aircraft is investigated. To allow the optimizer an investigation of the whole flight mission with a sufficient number of configurations, the individual calculations must be very fast and use only few variables. In more detailed investigations, not all disciplines are involved at the same time and perhaps not all mission points. The aerodynamic code can therefore use more time and variables to become more accurate. As a result, available turn around time, variables involved and accuracy achieved rise from step to step until the ultimate step of the flying aircraft.

As long as design modifications by theoretical predictions are relevant, turn around times are needed which allow for many repeated design loops. This is, depending on the step: one hour, one night, one weekend. Fast codes are all codes which allow for turn around times of one hour for pure aerodynamic calculations (with many individual code calls) or one night/ weekend for interdisciplinary tasks.

Very fast codes are closed formulas for the interdisciplinary investigations. They only need some main geometry parameters as input for global estimation of aircraft performance to allow configuration selection.

Fast codes relay on linearized theory with empirical corrections. They need more geometry parameters to allow for a first aerodynamic design optimisation including volume distribution and a first approximation of twist and camber; and they check the aerodynamic predictions in the interdisciplinary model.

Both codes calculate (at different accuracy levels) the global aerodynamic coefficients for performance calculations and first flight mechanics estimations. They identify the physical drag contributors and provide a load estimation.

As any code used for numerical optimisation, the codes must be robust. This means:

The code should be able to calculate all problems which the optimizer may pose. If the code breaks down, this must not stop the design process, but the code should deliver an inacceptably bad result which is the worse the heavier the code crash was. For instance, if negative pressures occur, the result can be a bad value proportional to the detected negative pressure value. This leads an optimizer to solutions, where the code does not crash. If the code is reliable, only those are interesting solutions. Such cases must be controlled by the design engineer!

Today, most research effort is devoted to highly sophisticated CFD-codes. These codes are needed and must be improved furthermore, but for a better and practical interdisciplinary aircraft optimisation, quality and applicability of the simple fast codes must be improved. Much more research effort is needed in this direction .

5.11 Accurate Computer Codes

Accurate computer codes are CFD-codes based on solutions of the Euler equations, sometimes with a coupled boundary layer solution and solutions of the Navier-Stokes equations. They are used for:

- configuration optimisation, to check and improve the previous design steps based on simpler codes, and to include wind tunnel results,

- interference drag reduction, which is impossible using simpler codes,

- inlet and nozzle design with strong shock/boundary layer interactions.

To allow efficient exploitation of CFD-codes, these codes must fulfill the following requirements:

- They must represent the relevant physical properties. For SCT design, these are:
  reliable radiation of disturbances (not fulfilled by most CFD-codes),
  prediction of shocks and shock reflections,
  prediction of separations (this still requires much research on turbulence).

- They must be able to use the exact geometry definitions including suited numerical grids.

- They must provide insight in flow physics by visualisation postprocessing of results.

- They must be able to predict aerodynamic loads.

- They must provide reliable performance predictions (drag prediction is still difficult for most CFD-codes).

- They must be able to identify the different physical contributions of drag (still a research task, especially for supersonic flow with strong radiation properties).

- They must provide reliable aerodynamic derivatives for flight mechanics calculations.

- They must support the analysis of experiments.

If these codes are only used to check results of previous calculations, the old fashioned procedure of man hour consuming grid adaption and numerical fine tuning may be applied. But as soon as the code is used for configuration optimisation, new requirements must be fulfilled:

- A geometry generator with very few variable parameters must model the variations of interest which the optimizer has to investigate.

- The grid generator must automatically provide a suited high quality grid.

- The code must fast and automatically converge to a useful result. If the code breaks down, a (bad) result must be provided which directs the optimizer to useful variations.

- The results produced by the optimizer's parameter variations must reflect the variation of physical results.

5.12 Inverse Design Capabilities

Since the introduction of direct numerical optimisation, importance of inverse design methods has decreased. Sometimes inverse design is seen as a relict of old design techniques. But inverse design remains important. There are cases, where the numerical effort of direct optimisation is still inacceptable. Though this will change in future, there remain other cases: Inverse design allows to construct solutions for comparison with incomplete or defective solutions. E.g. using only partial inputs or other than geometry inputs, a geometry can be designed to be compared with the geometry used for a CFD-calculation. Research is needed here, especially, if not only the classical inverse pressure design methods are to be used, but also other input alternatives.

5.13 Special Control Devices

An SCT has a flight envelope strongly enlarged in comparison to subsonic transports. All new configurations, either Concorde-like aircraft with thin wings or an OFW, may provide some configuration deficiencies unknown for subsonic transports. If the existing control devices cannot handle specific situations, or if the handling of those situations heavily penalizes those devices, then special control devices may have to be introduced for handling these situations.

Examples:

> If one engine stalls or an engine burst occurs at supersonic speed (OEI = One Engine Inoperative), strong lateral moments and rolling moments can establish. If handling of them penalizes rudder (and/or aileron) sizing, a special spoiler deflection on the other wing may compensate the occurring yaw, roll and pitching moments. Fine tuning is possible using the conventional controls.

> The oblique wing has superior aerodynamic performance, especially at low speeds. This may inhibit an acceptable landing procedure with steep descent. Special devices can produce the requested drag without inacceptable introduction of pitching moments.

Such devices strongly depend on the selected configuration. They are only recommended if they considerably reduce size, weight or complexity of the already existing system. It is possible, that such devices ease design of the control system layout, but complicating the system may occur as well.

5.14 Ejector Flaps

A new SCT, especially a Concorde-type SCT, has difficulties to fulfill take-off noise requirements. Any possibility to improve take-off performance and reduce noise must therefore be investigated.

The engine companies have proposed several engine types for SCT. There are engines which provide so much high pressure air at take-off, that they can only apply full (thermal) power if a large amount of bleed air is used elsewhere. It is worth, therefore, to investigate ejector flaps (Figure 34), mainly to increase thrust. Problems to be investigated are:

- What is the efficiency of the complex tubing and ejector flap system ?
- Does the additional installation weight of the complex tubing and flap system offset the improvement of take-off performance ?
- How complex and reliable will the system be ?
- What is the noise of such an ejector system ?
- Will exploitation of the ejector system for lift generation improve the design, when trim penalties and safety requirements are respected ?

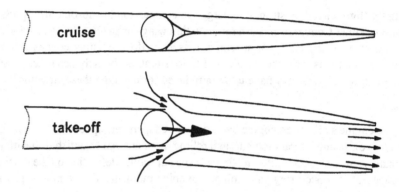

Figure 34   Ejector Flaps

To estimate the pros and cons of such a system, probably an SCT optimized without ejector flaps must be compared with a completely independent optimum design which is adapted to the exploitation of the ejector flaps.

5.15 Conclusion

A new SCT will only become reality, if many technologies are improved or newly developed. Some of them are aerodynamic technologies, as mentioned above. But many of them require contributions by other disciplines or need interdisciplinary connection with others. Both pure aerodynamics and interdisciplinary problems provide enough opportunities for many new intelligent contributions.

5.16 References

[58]   **Courant, R., Friedrichs, K. O.**
       Supersonic Flow and Shock Waves
       Applied mathematical sciences, 21, Springer Verlag, New York, Berlin, Heidelberg,
       1948/ 1976

[59]   **Courant, R., Hilbert, D.**
       Methoden der mathematischen Physik II
       Springer, 1968

[60]   **Jeffrey, A., Taniuti, T.**
       Non-linear Wave Propagation with Applications to Physics and Magnetohydrodynamics
       Academic Press, New York and London, 1964

[61]   **Mertens, J.**
       Instationäre Strömungen von Gasen mit brennbaren Partikeln, Kapitel A2
       Dissertation RWTH Aachen, 1983

[62]   **Sauer, R.**
       Nichtstationäre Probleme der Gasdynamik
       Springer, 1966

[63]   **Heinz, C.**
       Die Richtungs- und Verträglichkeitsbedingungen der Charakteristikentheorie
       Institutsbericht, Lehrstuhl für Mechanik, RWTH Aachen, 1970

[64]   **Küchemann, D.**
       The aerodynamic design of aircraft
       Pergamon Press, Oxford, 1978/ 1985

[65]   **Jones, R. T.**
       The Minimum Drag of Thin Wings at Supersonic Speed According to Kogan's Theory
       Theoret. Comput. Fluid Dynamics (1989) 1: 97-103

[66]   **Kogan, M. N.**
       On Bodies of Minimum Drag in Supersonic Gas Flow
       Prikl. Mat. Mekh., vol. XXI, no. 2, 1957, 207-212

[67]   **Mann, M. J., Carlson, H. W.**
       Aerodynamic Design of Supersonic Cruise Wings with a Calibrated Linearized Theory-
       Journal of Aircraft, **31**, 1, Jan.-Feb. 1994, pp. 35-40

[68]   **Van der Velden, A.**
       Aerodynamic Design and Synthesis of the Oblique Flying Wing Supersonic Transport
       PhD-thesis Stanford University, Dept. Aero Astro SUDDAR 621, Univ. Microfilms no.
       DA9234183, June 1992

[69]   *Engler, R. H., Hartmann, K., Schulze, B.*

Aerodynamic Assessment of an Optical Pressure Measurement System (OPMS) by Comparison with Conventional Pressure Measurements in a High Speed Wind Tunnel
Paper, presented at ICIASF '91, Washington D.D., 8 pages

[70]   Vollan, A., Alati, L.
A new Optical Pressure Measurement System
Paper, presented at ICIASF '91, Washington D. C., 7 pages

[71]   McLachlan, B. G., Bell, J. H., Park, H., Kennelly, R. A., Schreiner, J. A., Smith, S.C. Strong, J. M.
Pressure-Sensitive Paint Measurements on a Supersonic High-Sweep Oblique Wing Model
Journal of Aircraft, **32**, 2, March-April 1995, 217-227

[72]   Willert, C. E.
A Comparison of Several Particle Image Velocimetry Systems
DGLR-Bericht 94-04 "Strömungen mit Ablösung", Erlangen, 4.-7.10.1994, 266-271

[73]   Lienhart, H., Böhnert, T.
Grenzschichtmessungen an einem Laminarflügelprofil mit einem Laser-Doppler-Anemometer
DGLR-Bericht 92-07 "Strömungen mit Ablösung", Köln-Porz, 10.-12.11.1992, 471-476

[74]   Barreau, R., Renard, T. (Reporters)
BRITE EURAM Program "Supersonic Flow Phenomena", Final Report Subtask 1.3 "Preliminary Aeroelastic Investigation of Supersonic Transport Aircraft Configuration"

CHAPTER 6

CERTIFICATION OF SUPERSONIC CIVIL TRANSPORTS

J. Mertens

Daimler-Benz Aerospace Airbus GmbH, Bremen, Germany

6.1 Introduction

Since certification of Concorde new certification standards were introduced including many new regulations to improve flight safety. Most of these standards are to prevent severe accidents in the future which happened in the past (here: after Concorde's certification). A new SCT has to fulfill these standards, although Concorde had none of these accidents. But accidents - although they sometimes occurred only for a specific aircraft type - have to be avoided for any (new) aircraft. Because of existing aircraft without typical accident types having demonstrated their reliability, they are allowed to go on based on their old certification; although sometimes new rules prevent accident types which are not connected to specific aircraft types - like e.g. evacuation rules. Anyway, Concorde is allowed to fly based on its old certification, and hopefully in the future will fly as safely as in the past. But a new SCT has to fulfill updated rules like any other aircraft, and it has to be "just another aircraft" [75].

6.2 New Materials

A new SCT requires new materials to become viable - at least a new Concorde-type SCT; an Oblique Flying Wing (OFW) may possibly be using nowaday's materials. But before new materials can be counted on in aircraft design, *all* relevant material properties must be well known. And not only the material properties itself, but also the technologies to design for them, work with them, form them, join them with other parts of the same material or with the other materials used with them. When all this is known, we need the technologies and procedures to build and assemble large aircraft parts using these materials, which usually requires another set of manufacturing technologies. And last not least, we need certification rules and procedures applied to these new materials.

All these processes and the relevant data must be proven to be reliable and must be approved and certified by the authorities. This requires a long time, especially when fatigue related data are requested.

Before introducing a new material in aircraft design, it must be highly probable, that certification of the material and the related processes will be achieved timely. If this is not certain, a fall back solution must be developed (and possibly manufactured) in parallel! This is very expensive, consumes many resources and should be avoided whenever possible, but it can become necessary for risk minimisation in a tight time schedule.

6.3 Fatigue Testing of the Airframe

Before a new aircraft is allowed to transport passengers, it has to demonstrate its strength in a dynamic test. In this life cycle fatigue test, all important parts of the assembled test aircraft are loaded in a test rig with the load spectra which the aircraft has to expect in real airline operation. This includes the very rare maximum loads which are only felt by very few aircraft. To provide a safety margin, the fatigue test must simulate twice the cycle number (and load events) before final certification. Aircraft in airline operation must never reach more than half the number of cycles which were tested in the dynamic test rig.

For SCTs flying faster than Mach 1.8, this includes thermal fatigue testing and thermal cycling (like for Concorde). For SCTs flying at most Mach 1.8, the highest temperatures (about 80 °C) are reached on ground, when the aircraft stays in the sun without any wind. So these temperatures are not different to existing aircraft and well within the limits of known materials. But at Mach 2 mean temperatures at the aircraft skin are about 105 °C, the hottest points at 125 °C, and at higher Mach numbers even higher.

Thermal fatigue testing has to simulate the stresses related to temperature gradients. Temperature distribution and its gradients must therefore be simulated. Establishment of temperature gradients is a transient process which cannot be accelerated without changing (and mostly increasing) the gradients and stresses. Therefore cycling can only be accelerated for the equilibrium temperature times: long ground stops and cruise after heating up all fuel.

New SCTs will be designed for a service of 60 000 flight hours and 20 000 cycles. When assuming that the simulation time saved by equilibrium flight roughly is as long as the ground simulation time for cooling down, this means fatigue testing of 120 000 h or 13,7 years! Today's certification rules require certification to be completed within 12 years, although this may be altered for an SCT. Such long and complicated tests are very expensive. And faults detected during those tests may ground the whole fleet, at least the high cycle aircraft.

To save time and money it is proposed to develop a new technology for thermal fatigue certification: use partial simulation of representative subassembly parts with shorter heating and cooling cycles. These partial simulations will be transposed into a total aircraft simulation required for certification which is based on a certified numerical simulation method.

6.4 Noise Certification

For noise certification of an <u>aircraft</u> (with a specified engine type), the aircraft has to fly precisely defined procedures to enable quantified noise measurements. During those procedures, specific flight path properties are prescribed (which often are difficult to meet due to meteorological disturbances; this may require many test repetitions until the flight path requirements are met). Pilot actions are limited and must not include throttle or flap changes. Therefore, operational aircraft frequently are less noisy than certification flights.

In the future, standard procedures for flap and throttle scheduling may be accepted by the authorities for noise certification flights if those procedures are always performed automatically by the flight management system not requiring any pilot action.

Presently, no noise rules exist for new SCTs, which means, that without a change or extension of existing noise rules, new SCTs cannot be certified. It is expected, that future SCTs must comply with the rules system FAR 36, stage 3 (perhaps modified) or even a future stage 4.

Certification of <u>airports</u> usually includes noise exposure levels (noise pressure, weighed by a human sensitivity filter including a weight for time of day, integrated over the noise exposure time and the area) for the surroundings of the airports.

If the SCT will generate too much noise (e.g. in climb phases not certification-related), perhaps some airports will not accept SCTs although they can be certified according to the rules. This may happen because the *airport* can loose its certification, if noise exposure levels rise too much, due to too much or too noisy traffic. Possibly the airports have to select the less noisy traffic to maintain their concession.

But airport noise is not only related to aircraft (noise) performance, but also strongly to ATC (Air Traffic Control) procedures in the terminal area which often keep the aircraft for a longer time at low and noisy levels.

6.5 Emissions

Presently, emissions are respected only for the certification of engines with respect to a typical landing and take-off (LTO) cycle which is related to the airport area pollution.

Cruise emissions still are regarded to be (at least weakly) covered by LTO-requirements. Future will show if this assumption remains valid or if new rules will develop.

SCTs fly higher than subsonic aircraft; therefore their influence on the stratosphere and on the ozone layer is stronger. The stratosphere has not so much air exchange as the troposphere and therefore is more sensible. Interestingly, though, recent measurements - like the Airbus-program MOZAIC- indicate some strong global air exchange with the stratosphere.

Present research-based knowledge says that SCTs' influence on stratosphere and ozone layer is low, even below measurement accuracy ($\pm$ 1% steady state ozon change). SCTs below Mach 2 and subsonic aircraft mainly produce ozone, whereas SCTs beyond Mach 2 seem to destroy some ozone. Ozone production seems to change to ozone destruction at the upper limit of the tropopause at about 18 km altitude (a mean value, higher at the equator, lower at the poles and in winter). In troposphere and tropopause the NO_x of the aircraft exhausts generate ozone; but at least as important seems to be the strong affinity of NO_x to the gazeous HCl which is neutralized by NO_x in the mean latitudes, transported to the poles, stored there in polar stratospheric clouds (PSC) and later on released there. If this holds, aircraft are responsible for maintaining acceptable ozone levels in our mean latitudes by transferring some destruction to the poles.

Often water vapor in the exhaust is cited as a contributor to the green house effect. It is not yet clear which contribution is more important: the green house effect which reflects earth infrared energy back to the earth, or the contrail shielding which inhibits sun energy from reaching the earth's surface. In any case, the worldwide water vapor emissions of all aircraft have to be multiplied by 10^4 (!) in order to get a first very small reaction in the model calculations. On the other hand, phenomena of atmospheric physics are strongly nonlinear; so, today there exists not any useful indication on the importance of water vapor exhaust.

Altogether, scientific understanding and modelling of the stratosphere is not yet mature. Prediction tests of proposed changes mostly don't meet the data measured later-on, at least in the northern hemisphere with very complicated weather conditions. But a new SCT program will only be launched if certification and operation of many aircraft for many years can be based on reliable scientific predictions of the environmental implications.

6.6 Sudden Decompression

In the cabin of an aircraft, pressure is always held at comfortable levels corresponding to about 8 000 ft. This allows for proper function of the human biological system. If a hole in the pressure hull of an aircraft develops, not only oxygen must be provided to the passengers, but the aircraft has to descent immediately to an acceptabel altitude. Even when aspirating pure oxygen, it cannot be transferred to the blood at pressure levels comparable to more than about 35 000 ft altitude; and sudden decompression will cause the blood to boil after a short time.

In case of a pressure loss, therefore, the aircraft has to descend immediately and rapidly. During this steep descent, pressure levels in the cabin must never fall below cabin pressures of 40 000 ft; after at last two minutes a pressure altitude of at most 24 000 ft must be reached. To descend fastly without destruction of the aircraft, generation of drag is crucial (high altitude of about 50 000 ft must be reduced to 24 000 ft and simultaneously supersonic speed must be reduced to subsonic speed, both parts contributing almost the same part of required energy destruction).

There are two main reasons for sudden decompression:

- engine burst:
  probability of large holes can be reduced by suited positioning of the (most critical inner) engine(s).

- "20 ft^2 hole" (size may be a bit smaller for a narrow fuselage):
  size of this hole is pure geometrical. It was introduced after DC-10 accidents (which lost doors) and is mainly a door size. There is no measure to reduce probability or increase safety against this hole, e.g. by building double doors or a double hull; because geometry definition is not influenced by it. It cannot be seen now how this rule will be replaced by a physically based rule in the future.

At the high cruise levels of an SCT it is impossible to maintain the requested pressure levels using conventional techniques. Really new technologies and intelligent solutions are needed.

6.7 Controllability and Ride Comfort

SCTs are very drag sensitive. Control surfaces and stabilizer, are minimized therefore. Additionally, for slender configurations, the lever arms may become small. Consequently, unstable configurations are selected, but if stability may be achieved via artificial stabilisation, control authority against failure cases (like engine failure) and gusts or manoeuvers remains more critical. Many of the proposed configurations are unable to provide the required control authority, although - perhaps - they may achieve stability margins at the limit. Especially for some failure cases, like engine burst, missing redundancy of ailerons and elevons becomes critical for tailless configurations.

Long elastic fuselages (e.g. $l_o > 90$ m, 5 seat abreast) can easily provide loads of 1g for crew, front and tail passengers by fuselage oscillations during ground roll, take-off rotation and in turbulence - additionally to gust loads. This not only seems inacceptable, it really is inacceptable. Only very stiff materials are selected, weight seems to be at the limit; so a remedy seems only to be: wider, shorter fuselage with more drag and shorter range - or a completely different configuration, e.g. an OFW.

6.8 New Configuration Types

In the past, unknown configuration types mostly delivered some surprises during their flight tests: so called "unknowns", because new, unknown physical effects were detected, and "known unknowns", because physics were known in principle, but not respected because of the negligible influence of those effects on the older, known configurations. And we have to expect new "known and unknown" surprises with new configurations. Therefore, certification of new configurations and related technologies like

- Oblique Flying Wing aircraft (OFW)
- extremely flexible aircraft
- flutter sensitive aircraft
- artificial flutter supression
- supersonic laminar flow technology
- artificial view

probably will require demonstration of safe operation in a flying demonstrator. To *accelerate certification,* careful testing of a *demonstrator* aircraft with mature technology is absolutely necessary. To skip this step will most certainly turn out to be much more expensive and time consuming. (Although, there is a high probability that pennywise decision makers will not miss the opportunity to waste billions of $$...).

6.9 Conclusion

Development of a new viable SCT to become reality depends strongly on new technologies. To be applicable for civil aviation, these technologies must be certified. But even certification itself will require some special new certification technologies. To forget for this part will eventually cost additional time and money, although humans and (big) companies often prefer this route which delays the known necessary, but inconvenient decisions.

6.10 Reference

[75]   **Frantzen, C.**
Introduction to Regulatory Aspects of Supersonic Transports
Proceedings of the European Symposium on Future Supersonic and Hypersonic Transportation Systems, Strasbourg, November 6-8, 1989, paper II, 3.1

CHAPTER 7

GASDYNAMIC KNOWLEDGE BASE FOR HIGH SPEED FLOW MODELLING

H. Sobieczky

DLR German Aerospace Research Establishment, Gottingen, Germany

7.1 Introduction

This chapter is intended to illustrate a fragment of developments toward systematic high speed design, that is here aerodynamics in the regime of transonic and supersonic Mach numbers. The purpose is to show the modelling background of a combination of gasdynamics and geometry in the development of modern software for aerodynamic design in the virtual environment of personal and workstation computers. Here it is not intended to once more derive the basics for algorithm development in numerical simulation (CFD): only a simplified model of the basic equations is briefly mentioned because they paved the way to a better understanding of local flow phenomena, or as a consequence, of the requirements for detailed shaping of surface geometry in order to control local inviscid flow phenomena. In the transonic as well as in the supersonic regime, these phenomena are dominated by the interaction of surface geometry and surfaces within the flow field, for instance the boundary between locally subsonic and locally supersonic flow. These sonic surfaces, but also shock wave surfaces may be seen as part of the complete geometry set consisting of configuration and important flow features under design conditions. Motivation of this contribution is therefore to explain the gasdynamic background of some practical geometry tools for aerodynamic design, which take into account sonic and shock surfaces as part of the boundary conditions. Building on the pioneering basics of Guderley [76] and Oswatitsch [77], the ideas underlying the outlined concepts have been developed within the author's past theoretical work in transonics at DLR in Göttingen.

7.2 Control of Sonic Surfaces and Shock Waves

In a time long before the arrival of the digital computer, the model equations for compressible flow were derived. Since then, we know the Reynolds-averaged Navier-Stokes for the full problem, the Euler equations for their inviscid simplification and the Potential equation for a further simplification to isoenergetic flows. The latter extended the classical knowledge base of hydrodynamics into the compressible flow regime. A necessity to find solutions to these equations then led to several attempts to transform them, for instance to reduce the formidable difficulties stemming from the nonlinearity of the potential equation. In 2D flow, the hodograph transformation leads to an inversion of the problem, trading linear equations for nonlinear boundary conditions. Several mathematical methods were developed to create the first transonic airfoils. Elegant problem formulations could not hide the fact that solving mapped counterparts of real world problems never became very popular with the aerospace design engineer. Nevertheless, in a time when usually only numerical discretization of complex problems is seen as the way to get deeper insight into flow problems, some of these mapped model equations still have some value. One form of the "near sonic" model equations was found particularly useful, because it not only gave a number of flow models for transonic phenomena in closed analytical form but also led the way toward design principles for practical airfoils and wings.

7.2.1 Local 2D flow quality and singularity models

Near sonic Beltrami equations

The potential equation for a near-sonic plane or axisymmetric flow has a particularly elegant formulation in variables of state or in characteristic variables (for details see Ref [78]), illustrating the formal relationship between incompressible, transonic, plane and axially symmetric potential problems: characteristic equation and compatibility relation define a system of quasi-linear first order differential equations

$$V_t = Y^{P_1} U_s \tag{64}$$

$$V_s = jY^{P_1} U_t \tag{65}$$

$$X_s = U^{P_2} Y_t \tag{66}$$

$$X_t = jU^{P_2} Y_s \tag{67}$$

Here the variables (X, Y) denote the physical space and (U, V) a normalized pair of velocity variables, namely the Prandtl-Meyer function and the flow angle. Both pairs are dependent variables in a workspace (s, t) "Rheograph plane" which is identical with characteristic variables in supersonic flow (j=1, U>0) and their analytical continuation beyond the sonic line where j=-1, U < 0. Exponents p_1 and p_2 have a switch function: p_1 =0 denotes plane 2D flow, p_1 = 1 indicates axially symmetric flow; p_2 = 0 results in a simple mapping of linear subsonic or supersonic flow while p_2 = 1/3 switches to transonic flow. These equations include most of the flow models described by the pioneers in theoretical transonics, in closed analytical form or, for transonic axisymmetric flow where a weak nonlinearity persists, in a numerically very suitable form.

Equations (64)-(67) describe a large number of educational solutions which should be kept 'alive' as part of the knowledge base for transonic design and phenomena analysis. Because of (64)-(67) representing a system for quasi-conformal or characteristic mapping, solutions may be interpreted as transformations of geometries consisting of both the boundary conditions and all details of the flow, appearing as an analog flow in the rheograph plane.

Applications of mapped problems to test cases for numerical methods

In the transonic and low supersonic Mach number regime, analytical soloutions describing the local or asymptotic behavior of shock waves in the flow are of particular interest. Like the well-known logarithmic singularity [79] for the normal shock on a curved contour, there are other singular solutions for interaction of shock waves with solid boundaries or with flow phenomena near sonic conditions. Frequently, limiting cases between two known mathematical models have to be found. One such case is modelling a smooth transition from the mathematically well understood far field behavior of an airfoil in sonic flow $M_\infty = 1$ to the appearance of a detached shock wave in front of it if the free stream M_∞ slightly exceeds unity. This example was solved by a mapping to the rheograph plane [80].

More recently a similar problem was solved by a local mapping of the 2D plane flow near the tip of a wedge ramp at precisely the attachment Mach number $M_{\infty,att}$. The limiting case between a detached shock, normal to the flow axis, approaching the tip and an oblique shock attached to the wedge in supersonic flow (see the early experiments [81]), had to be found. Figure 1 illustrates this limiting case both in the physical plane and mapped to the rheograph plane. Results are given in [82], here it's just worth to mention that another, new logarithmic singularity at the wedge tip is found from simple mapping procedures.

The value of such local solutions for new and refined computational methods in fluid dynamics is evident: known local flow models represent exact solutions and refinements should give information about needed efforts to obtain such solutions numerically to a desired degree of accuracy. The wedge problem serves as a test case for an unstructured grid Euler code [83] where local grid refinements allow for a sharpening of the shock waves occurring in the flow, see Figure 36.

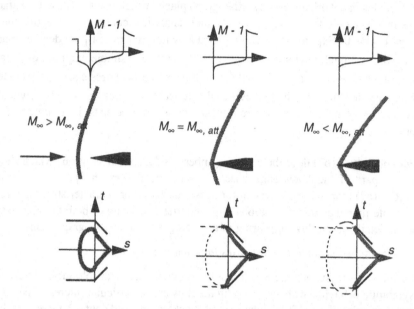

Figure 35   Understanding shock attachment / detachment to a wedge: Rheograph
mapping and shock relations transform problem to an analog flow detail and
conformal mapping case. From above: Axial and surface Mach number
distribution with changing wedge tip singularities; shock detachment,
limiting case and shock attachment; shock polar growing and intersecting
wedge boundary condition for increased Mach number.

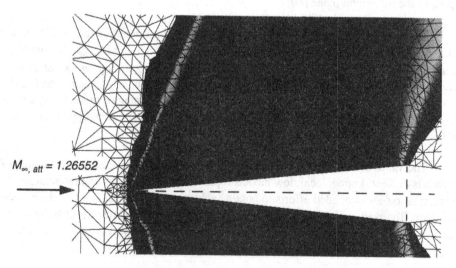

Figure 36   Numerical simulation of shock attaching to a wedge (slope 1:10). Simulation
with unstructured grid Euler code, grid adaptation near shock wave, isobars
visualization

7.2.2 Shock - sonic surface interaction in 3D flow

Compared to the rich variety of analytical models in plane transonic flow, our knowledge of local transonic phenomena in three-dimensional flow is much more limited. Equations (64)-(67) allow a construction of axisymmetric near sonic flow models, but the system is still weakly nonlinear so that basic solutions [78] may not be superimposed. Fully 3D flow does not permit a hodograph-type mapping, our knowledge base is restricted to experiments and evaluation of refined numerical analysis results. For design, a detailed knowledge base is needed to prepare a physically consistent and mathematically well-posed input for a reasonable numerical simulation. Inverse design methods are aimed at finding configurations with prescribed pressure distributions. This is a very attractive approach for practical design but inconsistencies between geometrical constraints and desired flow quality may occur: Ideal design goals may ignore the fact that unavoidable shock waves will accompany the envisioned flow patterns.

An example is the design of a swept transonic wing fixed to a wall or a fuselage without any fillet: the non-orthogonal corner angles between wing and sidewall do not allow for a smooth pressure distribution in the sonic expansion and recompression domain. Oblique shocks form and travel into the spanwise direction, eventually coalescing to stronger recompression waves, see Figure 37a. Only a locally unswept surface geometry can accomodate smooth expansion and recompression across sonic flow conditions (Figure 37b). Another way to avoid this purely inviscid feature of the flow and at the same time also influence viscous flow interaction favorably is the design of a fillet (Figure 37c) which provides a smooth surface without corners at the wing root. This qualitative sketch of 3D flow features is based on practical designers' experience, supported by an extrapolation of the well-developed knowledge base for 2D flows.

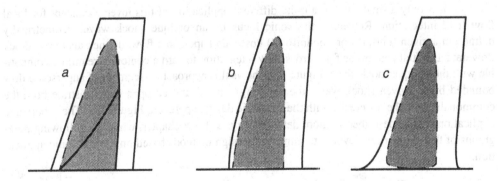

Figure 37   **Local supersonic domain (shaded) on wall-mounted swept wings in high subsonic Mach number flow: Oblique shock formation at wing root sonic expansion and recompression. Shock-free expansion and recompression requires locally unswept geometry (b) or fillet (c)**

7.3 From Flow Construction to Aerodynamic Design

7.3.1 Flow pattern integration from given sonic locus or shock

In the rheograph plane subsonic flow is separated from supersonic flow by the sonic locus, where $U(s,t) = 0$. Solving the model equations (64)-(67) for a transonic problem, where both types of flow ($U < 0$, $U > 0$) occur, therefore requires applying separate solution methods for both parts, with the need to have common data at the line $U(s,t) = 0$. For plane ($p_1 = 0$), transonic ($p_2 = 1/3$) flow, solution to (64)-(65) is the first step, because it is decoupled from (66)-(67). Choosing $U = s$, $V = t$ is just the simplest solution, without loosing generality of subsequently finding solutions $X = X(s,t)$, $Y = Y(s,t)$ to the linear system (66)-(67) in a second step. Creating a solution to both the subsonic part and the supersonic part now requires prescribing data $X^*(t)$, $Y^*(t)$ along $s = 0$. This data prescription is equivalent to defining the sonic line and a distribution of flow directions along it in physical space (X, Y). The data are used as part of the boundary conditions to an elliptic problem in the subsonic domain ($s \leq 0$) and as initial data for a hyperbolic problem in the supersonic domain ($s \geq 0$). Figure 38a shows the direction of integrating the local supersonic flow pattern: with the complete sonic line given, supersonic marching must be performed in a cross-flow direction, leaving the given sonic line unchanged and obtaining an arc of the surface geometry as a result. This way an inverse approach results in the goal of obtaining a surface geometry compatible with a strongly controlled sonic line. In the 1970's, techniques were developed to design transonic airfoils using this concept [84]. A rheoelectric 'analog computer' provided a very educational tool to understand also the background of Garabedian's method of complex characteristics to design shock-free airfoils [85], which was a mathematically elegant method but did not provide a lasting engineering knowledge base because of its complexity.

It is only a small step to a quite different application of this inverse concept for local flow field integration: Replacing the sonic locus by an oblique shock wave, geometrically defined in a given (uniform or nonuniform) upstream supersonic flow, defines also post shock flow data as initial conditions for a cross-flow integration toward a contour streamline compatible with the oblique shock wave. Figure 38b shows this approach: a segment of supersonic flow bounded by the given shock wave, the resulting contour and an open exit illustrate both the commonality and the difference with the transonic design approach, Figure 38a. The supersonic application will be described in more detail in the next book chapter, while the following paragraphs of this chapter are devoted to numerical design methods based on the transonic application.

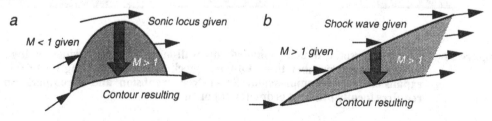

Figure 38   **Inverse design concepts in transonic and in supersonic flow.**

7.3.2 Characteristics in 2D and in 3D flow

In supersonic flow, Mach waves or characteristics are appearing as straight orthogonal lines ξ = t+s, η = t-s in the s,t rheograph plane. Basic system (1), or it's full compressible potential flow extension, see Ref. [86], maps these families of lines to 2D curves in the physical plane, within the domain of supersonic local Mach numbers. Figure 39 shows a 2D airfoil flow element with a chosen surface point C and both characteristics AC and CB intersecting it inclined to the flow direction with an angle α = arcsin(1/M): regions of influence from upstream and dependence to downstream are defined this way. For given flow data within the sonic line segment AB the solution for supersonic flow is completely determined within the triangle ABC. This is the basis for the 2D inverse method of characteristics [78] allowing the construction of a supersonic flow pattern.

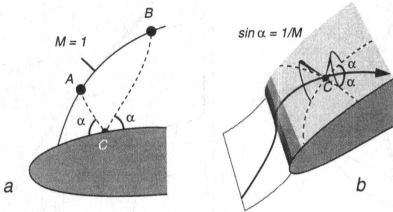

Figure 39   Characteristics (Mach waves) in transonic 2D flow (a) and 3D flow (b).

In 3D flow, both families of characteristics are Mach - conoid intersections, as illustrated in Figure 39b for a 3D wing element. They may intersect the sonic surface far away from surface location C and will not form closed domains comparable to the 2D sonic line interval AB. Mathematically an ill-posed problem, the inverse integration of 3D local supersonic flow fields with a potential flow cross marching technique nevertheless gave very satisfactory results for transonic wings [87], [88]. A more recent application for the design of supersonic wings, using an inverse version of the 3D Euler equations is outlined in the next book chapter.

Swept wings with subsonic leading edges

It is a well-known design principle to use two-dimensional airfoil flows in lower transonic or completely subsonic Mach number flows for the definition of swept wings in higher transonic or even supersonic Mach number flows. A shock-free transonic airfoil flow completely defines a family of 3D infinite wing flows yawed to upstream flow conditions with higher Mach numbers, with a relation between 2D and 3D Mach numbers $M_{3D} = M_{2D}/\cos(\lambda)$, where $M_{3D} = M_\infty$ for the wing and $M_{2D} = M_N$, the airfoil design Mach number and 3D Mach number component normal to the wing. Figure 40 shows an example, with characteristics evaluated for both 2D and 3D flow. The latter are 3D Mach conoid traces and mark regions of influence and dependence; a coalescence of characteristics in the location of 2D shock-free recompression illustrates the possibility of accumulative perturbation effects from distant spanwise locations in this area.

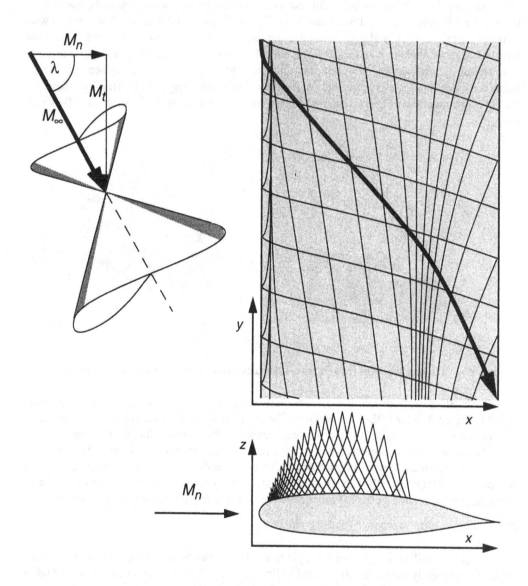

Figure 40   Using a 17% thick supercritical shock-free airfoil flow, ($M_n = 0.707$, $c_l = 0.6$) to define an infinite swept wing, $\lambda = 60°$, shock-free at $M_\infty = 1.414$, $C_L = 0.15$. Characteristics evaluation in supercritical airfoil flow and on swept wing upper surface.

7.3.3 Gasdynamic manipulations for design purposes.

The basic differential equations for compressible flow to be solved with numerical methods for practically interesting aerodynamic applications are the Reynolds-averaged Navier-Stokes equations. Neglecting viscous effects reduces this system of equations to the Euler equations. Gasdynamic phenomena such as shock waves in a flow field are simulated with sufficient accuracy by numerical algorithms for the latter inviscid model equations. Consequently, an inviscid flow design approach makes use of the Euler equations in suitable ways to help finding flows with reduced or vanishing shock waves. Techniques developed for shock wave control are then implemented successfully also in the Navier-Stokes equations. In this paragraph, a special approach to control shock wave strength is outlined: The Fictitious Gas (FG) method. This method was already developed when numerical algorithms to solve the compressible potential flow differential equations resulted in the first applied aerodynamic analysis computer programs [89]. In the meantime, the increased requirements for higher aerodynamic efficiency of an aircraft seems to justify a review of the theoretical potential of the FG method: In the following, first the purely "fictitious" part, i. e. the abstract mathematical model for the method is explained with the aim of a practical application. Second, the most recent results of classical shock-free flow construction by the FG method is illustrated and third, practical methods for rapid aerodynamic design derived from these approaches, using both the knowledge base for phenomena modelling and proven CFD codes is outlined.

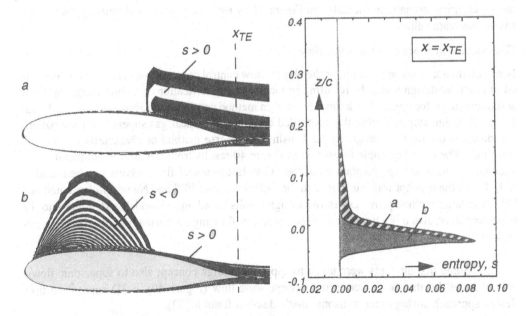

Figure 41　　Comparison of viscous transonic flow ($M_\infty = 0.7$, Re = 20 Mill., $c_l = 0.6$) past a given airfoil without (above, a) and with (below, b) flow control, carried out by energy removal within the supercritical domain. Entropy level isofringes and wake profile at the trailing edge indicate drag reduction of 35%.

Fictitious gas for a flow control concept

Born from originally a mathematically motivated manipulation to solve the mixed type differential equations modelling shock-free transonic flow, the definition of modified gas properties may be considered as a physical equivalent to this theoretical approach. Rheograph models for 2D flow for an analytical extension of the subsonic (elliptic type) problem into the supersonic domain in order to pose and solve a boundary value problem for shock-free airfoil flow results in a modified equation of state as part of the Euler or Navier Stokes equations. Solving such modified equations in direct 2D or 3D space for airfoils, wings or full aircraft configurations will therefore result in supercritical shock-free flow where the domain with velocities higher than the critical value is of a subsonic nature with locally "fictitious" properties, i. e. no practical realization of this flow is proposed so far.

Before this approach is explained as just the first part of a systematic and practical design method, it may be noted (Figure 41), that an interpretation of the fictitious gas as an ideal gas with pressure-controlled energy removal within the local supercritical domain lays ground for an interpretation as a thermodynamic flow control method for obtaining higher aerodynamic efficiency of given configurations. This Figure illustrates this for a thick wing section in transonic flow: Entropy contours and wake profiles at a control surface behind the airfoil allow to compare drag for these flows. Including viscous drag, aerodynamic efficiency (lift over drag) of the controlled airfoil flow is about 50% higher than for an uncontrolled flow. The remaining problem, of course, is the need to remove energy depending on the local pressure $p < p^*$ within the supercritical domain, as indicated in Figure 41 by the region with local entropy $s < 0$, relative to upstream values.

Construction of shock-free transonic flow

Not excluding a future application of the above 'flow control concept' we presently are interested in practical design methods, resulting in local shape modifications accomodating improved aerodynamic performance. The known FG design method combines the above flow control with a second design step replacing the controlled domain by an ideal gas supersonic flow pattern computed as outlined in paragraph 7.3.1, using the inverse method of characteristics or a 3D marching. The airfoil example illustrated in Figure 40 results from this design computation. A summary of various implementations of the FG to fast potential flow solvers can be found in [89]. Later, the concept was introduced to the Euler equations [90] and Navier Stokes equations [91]. The latter method provides viscous design results based on FG models extending into the boundary layer, also it is a time-accurate computation allowing for unsteady aerodynamic applications.

Some first attempts were taken to apply this design concept also to supersonic flows. As suggested by the shock-free infinite swept wing flow (Figure 40), a 3D supersonic flow design approach analog to the transonic method seems feasible [92].

Practical design methods for local shape modifications

During recent years a number of computational methods for designing 2D airfoils and 3D wings in the transonic regime have been developed. Among them are inverse methods, with the task to find airfoil or wing shapes for a prescribed pressure distribution. These latter methods are reviewed in chapter 10 of this book. The aim in the present chapter is only to illustrate some relations between gasdynamic properties of compressible flow and some practical consequences for 2D and 3D surface geometry quality.

In practical aerodynamics frequently a given configuration and its components may be modified only locally because of structural and other constraints. The FG design method is a practical tool if small local changes are allowed on the suction surface of a lifting wing: thickness distribution of a typical transport aircraft wing section may be reduced along a limited portion along chord less than half a percent to obtain improved aerodynamic efficiency up to 15 percent! Improvements also may be obtained for sections with local thickness added to the given geometry. To illustrate this, a series of simple design modifications to the thick airfoil of Figure 40 was carried out and is outlined in Figure 42 and Figure 43:

Following a FG computation (Figure 41), the method of characteristics defines a new section geometry on the upper surface (see the graphics of Figure 40). Contour differences are shown in Figure 42: given (baseline) wing section 1 and resulting section 2 differ by a Δz distribution, subtracted from the baseline. This Δz is smoothly shaped as a 'bump' function extending from the sonic expansion to the recompression location. A comparison of the aerodynamic properties for the baseline (1) and the new airfoil (2) is shown in Figure 43: Drag rise and airfoil polar obtained by CFD analysis [93] are illustrated. Related airfoil geometries with similar efficiency improvements may be obtained for mathematical approximations of the whole *subtracted* bump or its key features: amplitude and crest curvature. An airfoil '3' is obtained for a bump function with only smoothened (extended) start and end ramps, resulting in a slight loss in the Lift/Drag (L/D) ratio at design conditions, but with improved off-design performance, reducing 'drag creep' at lower Mach numbers. Airfoil '4' is obtained by *adding* a dual bump (now locally enlarging thickness everywhere along chord), modelling local curvature at x/c ~ 0.3 to be equivalent to the design modification. The resulting airfoil has still improved aerodynamic efficiency compared to the baseline. These latter *added* bump function may be combined with the concept of more local shape modifications like those used for influencing viscous interaction at the shock location [94].

There are various ways to calibrate geometrical models like the above used bump functions from observed phenomena like sonic surface geometries. The method of characteristics for 2D flow is the straightforward exact approach. Analysis of flow stream tube displacement requirements lead to even faster methods [95] which are suitable for 3D wing design.

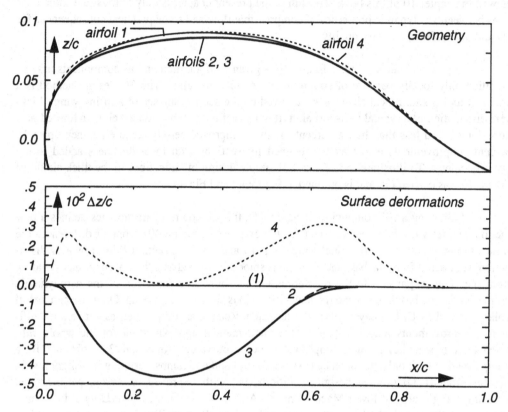

Figure 42   Upper surface of a thick baseline airfoil (1), shock-free design modification
(2), and relaxed geometry variations, for either optimum L/D (3) or
maintaining original thickness (4). Vertical coordinate differences Δz relative
to baseline airfoil.

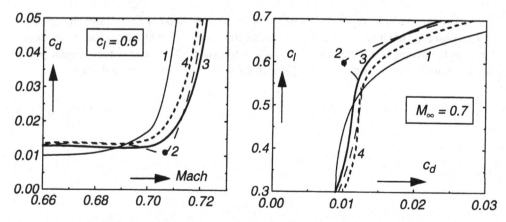

Figure 43   Drag rise for constant lift and drag polar at constant Mach number, for above
4 airfoils. Re = 20 Mill. (MSES [93] analysis)

7.4 Conclusion

In this chapter the attempt was made to illustrate some ideas which have been used and still seem to have innovative potential for the development of aerodynamic design tools. A refined analysis of inviscid flow phenomena is carried out by mapping into a suitable work space, resulting in mathematical models for local and asymptotic behavior of compressible flows. These models may be used for CFD code validation and also should be observed when prescribing target flow conditions in inverse design approaches.

Geometric boundary conditions, shock waves and the singularities connected with sonic flow conditions are shown to be strongly interacting, the knowledge about these phenomena may be used for aerodynamic design concepts. These lay ground for an inverse flow construction principle with applications to transonic as well as to supersonic flow. Numerical methods may be manipulated and recombined for obtaining airfoils and wings with improved aerodynamic efficiency and concepts for the development of novel flow control techniques could be encouraged by theoretical results.

Supersonic and hypersonic flow models derive from these principles, too, and will be used in the following book chapter for the design of advanced waveriders.

Numerical analysis and models to control gasdynamics result in a more detailed knowledge base about the shaping of flow boundaries, which translates into selecting a set of mathematical deformation functions for the optimization or adaptation of given baseline configurations. Being guided by the quality of the observed flow phenomena strongly helps to keep the number of control parameters for these functions small, as will be demonstrated in book chapter 9.

7.5 References

[76] **Guderley, K. G.**
Theorie Schallnaher Strömungen, Berlin, Göttingen, Heidelberg: Springer, 1957

[77] **Oswatitsch, K.**
Spezielgebiete der Gasdynamik. Wien, New York: Springer, 1977

[78] **Sobieczky, H., Qian, Y. J.**
Extended Mapping and Characteristics Techniques for Inverse Aerodynamic Design. Proc. Int. Conference on Inverse Design Concepts and Optimization in Engineering Sciences, (ICIDES III), Washington, D.C., 1991

[79] **Oswatitsch, K., Zierep, J.**
Das Problem des senkrechten Verdichtungsstosses an der gekrümmten Wand. ZAMM

40, pp 143-44, 1960

[80]   **Sobieczky, H.**
       Die Abgelöste Transsonische Kopfwelle. Zeitschr. Flugwiss. 22, No. 3, pp 66 - 73, 1974

[81]   **Liepmann, H. W., Roshko, A.**
       Elements of Gasdynamics. New York, London, Sidney: J. Wiley and Sons. pp 270 - 276,
       1957

[82]   **Sobieczky, H.**
       Anlegen der Kopfwelle bei Schallnaher Überschallanströmung. DLR IB 221-92 A 20,
       1992.

[83]   **Friedrich, O., Hempel, D., Meister, A., Sonar, Th.**
       Adaptive Computation of Unsteady Flow Fields with the DLR-TAU Code. AGARD
       Conf. Proc. CP 578, 1995

[84]   **Sobieczky, H.**
       Related Analytical, Anolog and Numerical Methods in Transonic Airfoil Design. AIAA
       79-1556, 1980

[85]   **Bauer, F., Garabedian, P., Korn, D.**
       Supercritical Wing Sections. Berlin, Heidelberg, New York: Springer, 1972

[86]   **Sobieczky, H.**
       Verfahren für die Entwurfsaerodynamik moderner Transportflugzeuge. (Aerodynamic
       Design Methods for Modern Transport Aircraft). DFVLR-FB-85-05, ESA Tech. Trans-
       lation ESA-TT-923, 1985

[87]   **Sobieczky, H., Fung, K-Y., Seebass A. R., Yu, N. J.**
       New Method for Designing Shock-free Transonic Configurations. AIAA Journal Vol.
       17, No. 7, pp. 722-729, 1979

[88]   **Fung, K-Y., Sobieczky, H., Seebass A. R.**
       Shock-free Wing Design. AIAA Journal Vol. 18, No. 10, pp. 1153-1158, 1980

[89]   **Sobieczky, H., Seebass, A. R.**
       Supercritical Airfoil and Wing Design. Ann. Rev. Fluid Mech. 16, pp. 337-63, 1984

[90]   **Li, P., Sobieczky, H.**
       Computation of Fictitious Gas Flow with Euler Equations. Acta Mechanica (Suppl.) 4:
       pp. 251-257, 1994

[91]   **Sobieczky, H., Geissler, W., Hannemann, M.**
       Numerical Tools for Unsteady Viscous Flow Control. Proc. 15th Int. Conf. on Num.
       Methods in Fluid Dynamics. Springer-Verlag: Berlin, Heidelberg, New York, 1997

[92]   **Li, P., Sobieczky, H., Seebass, A. R.**
       A New Design Method for Supersonic Transport. AIAA 95-1819, 1995

[93]   **Zores, R.**
       Transonic Airfoil Design using an Aerodynamic Expert System. AIAA 95-1818, 1995

[94]   Ashill, P. R., Fulker, J. L., Simmons, M., J., Gaudet, I. M.
A Review of Research at DRA on Active and Passive Control of Shock Waves. 20th ICAS Congress Conf. Proc. ICAS-96-2.1.4, 1996

[95]   Zhu, Z. Q., Sobieczky, H.
An Engineering Approach for Nearly Shock-free Wing Design. Chinese Journal of Aeronautics, Vol.2, No. 2, pp. 81 - 86, 1989

CHAPTER 8

CONFIGURATIONS WITH SPECIFIED SHOCK WAVES

H. Sobieczky

DLR German Aerospace Research Establishment, Gottingen, Germany

8.1 Introduction

Supersonic flow elements are valuable components for the development of new aerodynamic design concepts for efficient high speed flight vehicles. Among these elements, flows with oblique shocks may be used to shape socalled waveriders, which are almost classical cases where the component may already form a high lift-over-drag (L/D) configuration [96], [97]. In the past decades, the mathematical models of plane or axisymmetric supersonic flow fields with shocks have been used to create a number of simple test cases for experimental investigation, long before numerical flow analysis methods and large computers were available.

Today we have various inviscid and viscous CFD analysis methods operational to investigate the aerodynamic performance of airfoils, wings and 3D configurations in design and off-design conditions. Waveriders are ideal test cases for numerical methods simulating inviscid compressible flow: Special known shock patterns occur on relatively simple geometries at design conditions. In this situation we may think about creating more general and perhaps more efficient waveriders by new design methods, mature CFD analysis will verify the predicted inviscid flows and help to analyze design and off-design conditions performance, allowing for substantial reductions in systematic windtunnel testing.

Inverse aerodynamic design is an approach to obtain configuration elements compatible with certain desired performance characteristics. In supersonic and hypersonic flow, the shock wave formed by the lifting body at the leading edge is carrying much of the information *about lift, wave drag and noise*, which are key issues in applied supersonic aerodynamics. It seems a challenging task therefore to invert the design problem by controlling the bow shock wave and find compatible body surfaces. Since the 1960's waverider configurations are constructed from simple known plane or axisymmetric flow fields with oblique shocks, suitably cut by stream surfaces to provide special delta wings with sharp leading edges.

In this chapter waverider flows are generated from given shock wave geometries. In the past years since Nonweiler's first caret wing waveriders [98], plane 2D and axisymmetric conical flows have been used for the generation of lifting wings in supersonic flow, with the detailed inviscid flow structure coming as a result with the geometric shape. In the recent years a revived interest [99] in these waveriders has emerged, to use them as baseline configurations with known aerodynamic performance at idealized design conditions, for further development and optimization using CFD as well as refined experimental techniques. Two new concepts [100] were presented for this purpose: 'Osculating Cones' and 'Inverse Euler Marching' techniques, to obtain more general waverider geometries than derived previously from plane or conical generating flows. These concepts and applied work based on them are reported here in the following.

8.2 Construction of Flows with Oblique Shocks

This work is an approach to compute supersonic flowfield models which are mathematically simulating a part of the inviscid flow past an aerodynamic configuration. For waveriders, these flows will be bounded by a shock wave, a stream surface defining a solid body contour, and an exit surface. Such boundaries allow to place this model within given supersonic upstream flow conditions, connected along the shock by the Rankine - Hugoniot relations.

8.2.1 Cross - marching in supersonic flow

An earlier, similar approach, as described in the previous book chapter, was aimed at connecting a local supersonic flow field with surrounding subsonic flow along a prescribed sonic line. This procedure was used to model transonic flows with applications in supercritical wing-design. This process was first carried out as a linear method of characteristics, it is equivalent to marching in a direction normal to the flow ("cross - marching"), thus allowing for a start at initial conditions compatible with embedded supersonic flow domains.

Locally axisymmetric flow

Within 2D inviscid supersonic flow, a local linearization of the basic equation for a velocity potential Φ gives the wave equation

$$\Phi_{xx} - \Phi_{zz} = 0 \tag{68}$$

where x is the direction of the local flow vector. We realize that both x and z are time-like directions, a marching in the flow direction or normal to it are mathematically equivalent, their choice determined only by formulation of the initial conditions, see Figure 44.

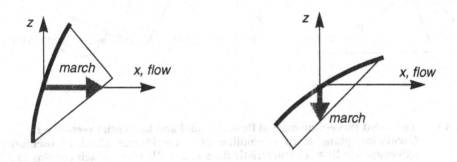

Figure 44   Initial conditions for downstream marching (a), cross - marching (b)

Axisymmetric flows are a special case of 3D flows, but their computation may also be carried out in a 2D meridional plane by the method of characteristics. It can be shown that the axisymmetric model equation for linear supersonic flow

$$\Phi_{xx} - \Phi_{rr} - \Phi_r / r = 0 \tag{69}$$

can be used to second order approximate locally a 3D flow element which is a solution to the general 3D flow equation

$$\Phi_{xx} - \Phi_{yy} - \Phi_{zz} = 0 \tag{70}$$

Location of the axis of this osculating axisymmetric flow depends on local flow curvature and velocity gradient. This can be used to develop a 3D method of characteristics, which has locally 2D properties and therefore reduces perturbation amplification in a numerical cross - marching approach.

Figure 45 illustrates 2D flow and the possibility to locally approximate 3D flow by an axisymmetric flow model.

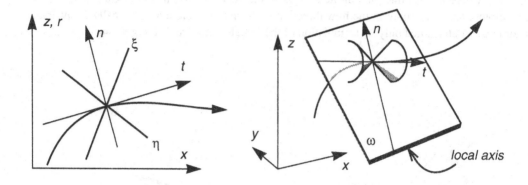

Figure 45   **Two - and three-dimensional flow. Normal and tangential vectors n,t. Osculating plane ω to streamline as a meridional plane of osculating axisymmetric flow, characteristic lines x, h in 2D flow, Mach conoids in 3D flow.**

The abovementioned idea of cross - marching is intended to familiarize the reader with a strategy of numerically integratung flow model equations from initial data determined solely by unperturbed upstream flow and a geometrically defined oblique shock wave. The flow field behind this shock has to be evaluated by using the method of characteristics, for irrotational flow equivalent to a potential flow solver, for rotational flow resulting from an arbitrarily curved shock wave equivalent to an Euler solver. Boundaries given for and resulting from characteristics calculation in the following will be explained for the simple flow past a wedge, because using a marching code later will require to keep in mind regions of dependence if a numerical scheme should work properly.

The flow past a wedge is sketched in Figure 46 to show how cross - marching will be used to obtain the flow field solution in part or in whole. Let supersonic flow be deflected by a wedge contour AG. A shock AB forms end bends the streamlines within triangle AGB conformal to the wedge angleWe ask now for the inverse computational approach to find the flow and the contour behind the given portion AB of the shock wave. Characteristics resulting from cross - marching will define a triangular region of dependence ABC, which includes a non-physical part of the solution beyond the contour AD, which results from flow field integration within ABC. On the other hand , a part DBG of the physical solution is not available with initial data given only along AB. Data need to be given also along a portion BE of the exit, with results along CB available this defines the solution within the polygon CBEF. Evaluation will define the streamline continuation beyond D toward an exit value at Gsituated on BE or EF, depending on the choice of E. For high supersonic Mach numbers, contour length AD may be only a fraction of the continuation DG.

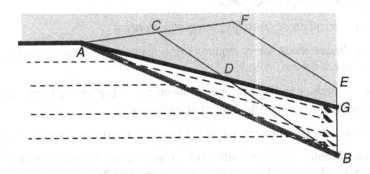

Figure 46   Supersonic flow past a wedge: Regions of dependence for cross - marching, with given shock AB and given exit BE.

For curved shock waves prescribed, we have to expect rotational post-shock flow resulting from the computation. The contour streamline is another characteristic now, cross - marching requires avoiding contradictory initial data along AB and BE, if both domains have to be solved. Furthermore, data for entropy or vorticity distribution are not defined beyond the contour (to be computed), which may pose a problem for cross - marching.

Another feature of cross - marching with the method of characteristics is the possibility of limit lines occurring in the computed flow. A multivalued solution may be found because the marching is essentially carried out in a hodograph plane: i. e. a variable of state, for instance the velocity q , is the independent variable to march along, while coordinates x of physical space are resulting, see Figure 47. Such a solution cannot be obtained by marching in physical space: the marching direction would have to be reversed to pick up a continuous solution $q(x)$. Occurrence of a limit line or surface alerts us that the given shock wave is not compatible with a smooth flow downstream of it, the initial data should be changed.

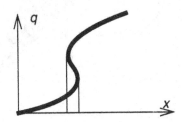

Figure 47   Representative variable of state q as a function of representative space coordinate x: multivalued solution if limit lines (surfaces) occur.

We draw some conclusions from this outline of models to be simulated numerically and used for practical design aerodynamics:

Caveats for a numerical cross - marching integration concept

1.Given oblique shock waves require cross - marching to obtain a contour compatible with initial conditions.

2. Cross - marching in 2D plane or axisymmetric flow is effectively carried out by well - posed hodograph (inverse) methods of characteristics.

3. 3D flows may be approximated locally by (osculating) axisymmetric flow.

4. Extent of contours designed with cross - marching may be small compared to extent of the given shock wave in high Mach number supersonic flow.

5. Rotational flow constrains region of dependence to physical flow bounded by designed contour.

6. Occurence of limit surfaces requires initial data modification; inverse marching allows to pick up limit lines, direct space marching results in infinite gradients.

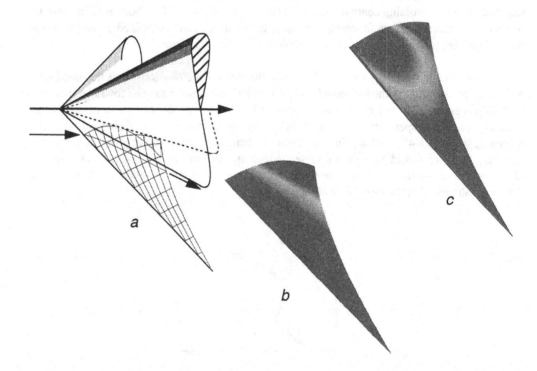

Figure 48   **Mach number $M_\infty = 2$, circular cone flow integration and flow element past a curved shock wave. Given shock cone angle 45° (a), resulting solid cone angle 27.32° and analytical continuation toward limit cone (18.5°) within solid cone boundary. Isomach fringes of conical (b) and curved shock (c) flow field .**

8.2.2 Axisymmetric flows with given shock waves

Examples showing the possibilities to compute axisymmetric flow elements resulting from given shock waves are given in [101]. The well known flow past a circular cone in supersonic flow is verified as well as generalizations for curved shocks with varying strength followed by rotational flow. Figure 5 illustrates the solution for a cone flow and for the flow past a curved axisymmetric shock wave element, as it occurs in the flow past an ogive body nose. The characteristics grid depicts the full solution extending beyond the contour streamline: a limiting cone is found with no solution appearing near the cone axis.

This fast numerical method of characteristics is an inverse Euler solver to design 2D and axisymmetric flows with given shock waves. The flow past a circular cone has been described by Taylor and Maccoll , its conicity reduces this special case of axisymmetric flow to solving one ordinary differential equation. This will be used in the following description of practical design methods by locally applying conical flow as an 'osculating flow pattern' in more general 3D boundary conditions. Curved shocks will allow for an even wider variety of using an axisymmetric flow element for 3D flow design.

8.3 Osculating Cones (OC) Waveriders

8.3.1 Generalization of the conical flow waverider

A suitable selection of streamsurfaces from a given analytical or numerical solution for a 3D supersonic flow with a shock system may define the solid boundary of an obstacle in this flow, with the shock waves inthe flow generated by this obstacle. This is the basic design principle of the waverider, which is basically a simple supersonic generic aircraft. Renewed interest in high Mach number supersonic and hypersonic transport vehicle design revived the classical waverider principle using wedge and cone flows to carve out a multiplicity of shapes, suitably parameterized and shapes selected by optimization strategies [102]. This is made possible in a fast and efficient way because the used flow solution is known and so are pressures along the contour, integrated to lift and drag at inviscid flow design conditions very easily. Furthermore, known streamlines allow for a quasi - 2D boundary layer computation, subsequent shape correction and viscous drag computation. Viscous flow optimized waveriders are therefore the starting point to a refined configuration definition for high speed transport aircraft.

Combining plane and axisymmetric flow: Slope surface shock waves

Waverider aircraft have supersonic leading edges of a completely integrated wing - body configuration. Planar 2D flows are suitable for lifting wings with sufficient span but suffer from low volumetric efficiency. Conical flows give better volume but have a small aspect ratio resulting in bad low speed performance. A compromise is a higher order approximation to exact model flows by a spanwise variation of the local cone radius: it is an application of the above mentioned validity of *locally* axisymmetric flow. Figure 49 illustrates the geometric definition of a conical or generalized shock segment bounded by the waverider inlet capture curve (ICC) coordinates and the leading edge which serves as a flow capture curve (FCC). This definition of a shock wave patch with given obliquity angle β includes planar, conical and slope surface shock waves with constant strength. The latter are envelopes of a single cone sliding along the given leading edge (FCC) in 3D space, and the idea to approximate the resulting flow field behind such shock waves by the local tangent cone flow meridional (osculating) plane results in relatively high accuracy of this straightforward definition of the lower surface along with its pressure distribution.

As in previous waverider work, the upper surface may be a free stream surface or has to be designed for additional lift by convex contouring for flow expansion. In the following illustrations some examples designed with software based in this principle is shown. Fast design codes based on this method have been developed by various authors. Extensions for refined viscous flow modelling and for incorporating propulsive devices have been developed.

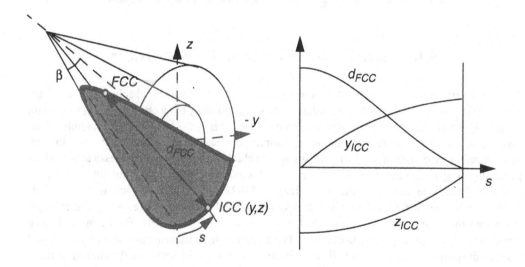

Figure 49   Defining the bow wave of a conical flow waverider: Inlet capture curve (ICC) and leading edge or flow capture curve (FCC). Generalizations for OC waveriders if ICC is not a circle, shock segment is a slope surface, shock angle is β

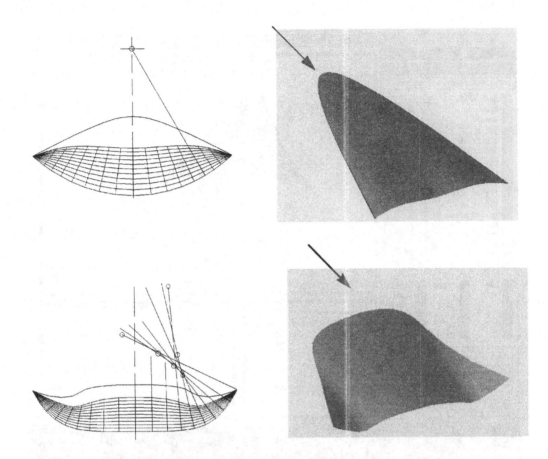

Figure 50   **Waveriders generated from bow shocks (shock angle $\beta = 18°$) in supersonic freestream of $M_\infty = 4$. Cone-derived (above) and OC configuration**

8.3.2 Design and optimization software based on the OC method

The OC method has proven to be a very flexible design tool requiring only a few input parameters for obtaining a wide variety of configurations with supersonic leading edges, plus (and that's the attractive feature) it also gives results for the complete flow field and lift and wave drag of the configuration. In this situation a flexible definition of the 3 functions (y_{ICC}, z_{ICC}, and d_{FCC}) of ICC arc length s, see Figure 49) is most important for rapid pre-design and optimization studies. The first application software illustrated below combines the OC method with geometry input used in a more general configuration generator outlined in the next book chapter. The latter makes possible a definition of curves with piecewise analytic structure and control of first to third derivatives which are needed for the calculation of local cone axes as depicted in Figure 50.

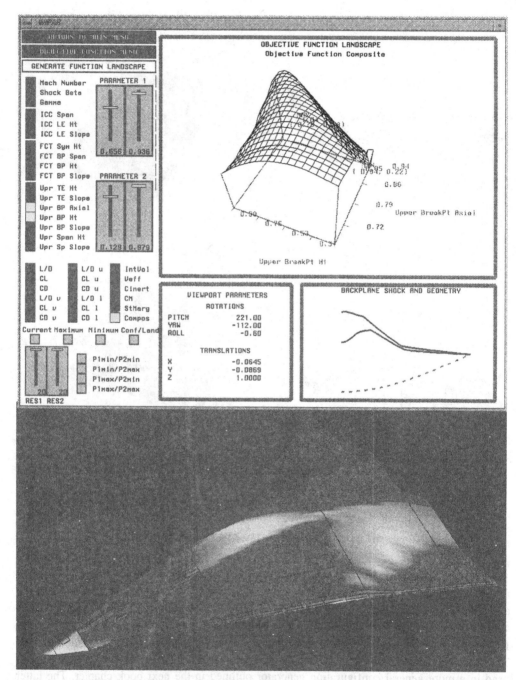

Figure 51   WIPAR Waverider design results: Window environment for L/D optimization (above), resulting optimized, animated waverider configuration for a TSTO mission (below)

The WIPAR interactive code

A versatile computational tool, the "Waverider Interactive Parameter Adjustment Routine - WIPAR" - [103], [104]was written for usage on graphic workstations, allowing for a fast and educational display of the resulting configurations and aerodynamic coefficients at design conditions. Some features of this powerful software are illustrated in Figure 51: The windows environment with various adjustment sliders, quick look diagrams and key results displayed make this computer code a suitable baseline for a hypersonic configuration design expert system. Various configuration examples have been studied and some have been manufactured for experimental investigation in wind tunnels. Experiments in the subsonic Mach number regime have been carried out for a waverider similar to the configuration depicted in Figure 51, which was optimized for Mach 4. Results confirm conclusions that waveriders based on the OC concept may also have quite favorable aerodynamic characteristics in the low speed regime [105].

Aerospace vehicle and SCT aircraft optimization studies based on WIPAR

The WIPAR software is used for an implementation in practical aircraft and aerospace vehicle design systems. For improved modelling of viscous flows and thermal loads, refined methods for boundary layer and heat transfer have been developed and coupled with the design program [106], [107].

With the main thrust of waverider applications so far being directed toward aerospace vehicle operation at hypersonic Mach numbers, an alternative use is proposed for the layout of aircraft wings for high speed transport at even the lower supersonic Mach numbers [108]. Here the high Mach number design cases are found to have remarkably good L/D ratios and are therefore investigated further, namely regarding the influence of sweep on the occurrence of boundary layer instabilities and the potential of such wings for laminar flow control. Flexibility in creating configurations for missions in the supersonic as well as in the hypersonic flight regime makes the waverider concept attractive for a variety of high speed transport concepts. Investigations on a series of configurations derived from WIPAR generated shapes have already been studied for TSTO missions using a multidisciplinary aircraft design program [109].

Propulsion integration on OC waveriders

The flow field including streamlines between the lower surface of a waverider and the oblique shock is readily available for OC waveriders at design conditions. This may be used for adding stream-aligned surfaces like the propulsion casing, and furthermore, flow conditions at the inlet capture curve ("ICC") may be input for internal flow simulation of the propulsion unit. This way generic waverider configurations with a flexibility obtained using the OC concept are ideal input for design considerations beyond purely aerodynamic aspects: Multidisciplinary design optimization of realistic vehicles will require a synchronized development of aerodynamic, structural, aeroelastic and propulsive components, among many other considerations. Some aspects are already studied in the design of waveriders with integrated propulsion including inlets and nozzles [110].

8.4 Waveriders with Arbitrary Shock Waves

The OC waverider design method has proven to be extremely useful for practical design tasks so far, but further generalizations seem worthwhile for curved bow shock waves: a replacement of conical flow by an axisymmetric base solution resulting from the 2D inverse method of characteristics may give waveriders with better volumetric efficiency because the convex shock shapes to be prescribed as illustrated in Figure 48 (c) result rather from blunter ogive - shaped forebodies than from cones.

Inverse solution to the 3D Euler equations with the SCIEMAP code

A 3D arbitrary but analytically controlled oblique shock wave requires an even more general approach: a marching procedure toward the contour surface to be found has to be applied. An inverse solution to the Euler equations results from such an approach. After some first results [100] confirming the concept, a numerically more advanced treatment taking into account the '*caveats*' postulated in a previous paragraph led to an inverse Euler solver 'SCIEMAP - Supersonic Cross-stream Inverse Euler MArching Program' [111].

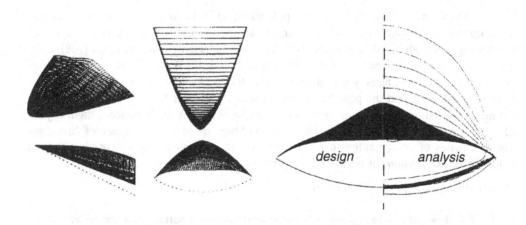

Figure 52   **SCIEMAP - designed Waverider with variable strength shock wave; CFD inviscid analysis (code: F3D), verification of the shock wave geometry.**

The 5 equations of motion (continuity, energy and 3 momentum equations) plus 10 equations for the expansion of the 3D flow gradients on the post-shock wave front with its given geometry and 5 primitive variables resulting from the Rankine - Hugoniot oblique shock relations give a system of equations for the 3D gradient vector of the primitive variables. This is suitably used in a local domain marching grid observing the constraints and limitations given by the *caveats*. Some promising configuration studies have been carried out [111], [112]; an example is illustrated in Figure 52. This method has the potential to be applied to other problems where an inverse solution to the compressible steady flow Euler equations seems promising.

8.5 Conclusion

Two inverse design methods have been presented here to obtain supersonic and hypersonic waverider shapes with geometries more general than available from previously used known flow fields: The Osculating Cones (OC) concept is realized in the WIPAR computer code and the inverse integration of the 3D Euler equations from prescribed shock surfaces is performed with the SCIEMAP code.

In the first approach use is made of axisymmetric flows to construct flows with more general shock wave shapes (slope surfaces), but still constant shock strength. The WIPAR software is an interactive computer program for graphic workstations and allows for a flexible parameterized input of the basic aerodynamic and geometric parameters of waverider wings, like the design Mach number, shock angle and planform geometry. Applications have been carried out adding extensions for viscous flow and thermal load. Case studies making use of the OC concept serve as test cases for optimization strategies and propulsion integration.

In the inverse Euler design approach, cross-marching is an exact method of characteristics for inverse 2D and axisymmetric calculations but mathematically ill-posed in general 3D flow modelling; nevertheless it is used successfully to create waverider shapes with more general shock shapes of varying strength. The new numerical marching technique makes use of some features of characteristic cross-marching, results were obtained confirming shapes known from other methods and also completely new with improved volumetric efficiency.

Comparing both inverse methods, each approach has its merit: While SCIEMAP may be used for more general fluid mechanic modelling requiring solution to an inverse problem and the waveriders created may have better volumetric efficiency, the use of the WIPAR software enables the designer to comfortably create shapes containing various geometric details similar to a configuration generator for general purpose, as being outlined in the next book chapter.

8.6 References

[96]   **Kuechemann, D.**
       The Aerodynamic Design of Aircraft. Oxford: Pergamon Press, 1978

[97]   **Anderson, J. D.**
       "Hypersonic and High Temperature Gas Dynamics", New York: McGraw Hill Book Company, 1989

[98] **Nonweiler, T. R. F.**
Delta Wings of Shape Amenable to Exact Shock Wave Theory. J. Royal Aeron. Soc. Vol 67. No. 39 1963

[99] **Anderson, J. D., (Editor)**
First International Hypersonic Waverider Symposium, University of Maryland, 1990

[100] **Sobieczky, H., Dougherty, F. C., Jones, K. D.**
Hypersonic Waverider Design from Given Shock Waves. Proc. First Int. Hypersonic Waverider Symposium, University of Maryland, 1990

[101] **Sobieczky, H., Qian, Y. J.**
Extended Mapping and Characteristics Techniques for Inverse Aerodynamic Design. Proc. Int. Conference on Inverse Design Concepts and Optimization in Engineering Sciences, (ICIDES III), Washington, D.C., 1991

[102] **Bowcutt, K., G., Anderson, J. D., Capriotti, D.**
Viscous Optimized Hypersonic Waveriders. AIAA paper 87-0272, 1987

[103] **Center, K. B.**
An Interactive Approach to the Design and Optimization of Practical Hypersonic Waveriders. Ph. D. Thesis, University of Colorado, Boulder, 1993

[104] **Center, K. B., Sobieczky, H., Seebass, A. R., Dougherty, F. C.**
A Strategy for the Design of Non-conical Waveriders from Arbitrary Shock Profiles. J. Aircraft, (submitted for publication)

[105] **Miller, R. W., Argrow, B. M.**
Subsonic Aerodynamics of an Osculating Cones Waverider. AIAA-97-0189, 1997

[106] **Eggers, Th., Sobieczky, H., Center, K. B.**
Design of Advanced Waveriders with High Aerodynamic Efficiency. AIAA 93-5141, 1993

[107] **Streit, Th., Martin, S., Eggers, Th.**
Approximate Heat Transfer Methods for Hypersonic Flow in Comparison with Results Provided by Numerical Navier-Stokes Solutions. DLR FB 94-36, 1994

[108] **Radespiel, R., Stilla, J., Eggers, Th., Sobieczky, H.**
Aerodynamic Design of Supersonic Wings with High Aerodynamic Efficiency, Proc. 7th European Aerospace Congress, Toulouse 25-27 Oct. '94, 1994

[109] **Strohmeyer, D., Eggers, Th., Heinze, W., Bardenhagen, A.**
Planform Effects on the Aerodynamics of Waveriders for TSTO Missions. AIAA 96-4544, 1996

[110] **Takashima, N., Lewis, M.**
Engine-Airframe Integration on Osculating Cone Waverider-Based Vehicle Designs. AIAA 96-2551, 1996

[111] **Jones, K. D.**
A New Inverse Method for Generating High-speed Aerodynamic Flows with Applica-

tion to Waverider Design, Ph. D. Thesis, University of Colorado, Boulder, 1993

[112] **Jones, K. D., Sobieczky, H., Seebass, A. R., Dougherty, F. C.**
Waverider Design for Generalized Shock Geometries. J. Spacecraft and Rockets Vol. 32, No. 6 pp. 957 - 963, 1995

CHAPTER 9

GEOMETRY GENERATOR FOR CFD AND APPLIED AERODYNAMICS

H. Sobieczky

DLR German Aerospace Research Establishment, Gottingen, Germany

9.1 Introduction

This chapter is intended to combine the knowledge bases of applied geometry with those of hydrodynamics and aerodynamics, including the modest additions presented in the two previous book chapters focusing on the interaction between compressible flow with shock waves and flow boundary conditions. The need to have flexible tools for effectively influencing the phenomena occurring in high speed flow calls for development of fast and flexible software to create shapes in a way to have easy access to the crucial shape-generating parameters controlling these flow phenomena, and at the same time observe the constraints given by structural and other practical limitations.

Renewed interest in Supersonic Civil Transport (SCT) or High Speed Civil Transport (HSCT) calls for extensive computational simulation of nearly every aspect of design and development in the whole system. CAD methods are available presently for many applications in the design phase. Nevertheless, work in early aerodynamic design lacks computational tools which enable the engineer to perform quick comparative calculations with gradually varying configurations or their components. To perform aerodynamic optimization, surface modelling is needed which allows parametric variations of wing sections, planforms, leading and trailing edges, camber, twist and control surfaces, to mention only the wing. The same is true for fuselage, empennage, engines and integration of these components. This can be supported in principle by modern Computer Aided Design (CAD) methods, but data preprocessing for numerical flow simulation (CFD) calls for more directly coupled software which should be handled interactively by the designer observing computational results quickly and thus enabling him to develop his own intuition for the relative importance of the several used and varied shape parameters. The requirements of transonic aerodynamics for transport aircraft in the high subsonic flight regime as well as more recent activities in generic hypersonics for aerospace plane design concepts have enhanced previous activities [113], [114] in the development of dedicated geometry generation [115]. Based on experience with the definition of test cases for transonic aerodynamics [116] and with fast optimization tools for hypersonic configurations outlined in the previous chapter, as well as taking into account new developments in interactive graphics, some fast and efficient software tools for aerodynamic shape design are already operational or under development. The concept seems well suited for application to various design tasks in high speed aerodynamics and fluid mechanics of SCT aircraft projects, especially with options to select suitable parameters for an application of optimization strategies which will be presented in following book chapters.

It is the author's intention to illustrate the options of the proposed method for a systematical development of some of the required technologies for high speed aircraft design, at least those needed in aerodynamics, some for aeroelastics and for aeroacoustics. Computational simulations will have an ever increasing share in technology development though experiments are still needed; wind tunnel models are to be created by CAD systems for which the geometry generator as a preprocessor must provide data of exactly the same accuracy as for CFD.

Much use is made of graphic illustrations in this chapter which is natural for this topic and which may be more useful than much text. A powerful interactive fluid mechanics visualization software system [117] greatly adds to an efficient use of shape design methods: structured and unstructured CFD grids, shaded solid surfaces and isofringes depicting flow variables distribution results are displayed on a graphic workstation screen and for a few examples in the following pages.

9.2 Parameterized Curves and Surfaces

The geometry tool explained here has been developed in the years shortly before interactive graphic workstations became available, originally for input with data lists but increasingly laid out for interactive usage in the windows environment of the workstation. The list input still is the basic option and data for such usage will be presented here for explanation. Focusing on surface modelling of aerodynamically efficient aircraft components, we realize that the goal of shape generation requires much control over contour quality like slopes and curvature, while structural constraints require also corners, flat parts and other compromises against otherwise idealized shapes. When familiarity is gained with a set of simple analytic functions and the possibility is used to occasionally extend the existing collection of 1D functions, ground is laid to compose these functions suitably to yield complex 2D curves and finally surfaces in 3D space. This way we intend to develop tools to define data for airframe components with a nearly unlimited variety within conventional, new and exotic configurations. A brief illustration of the principle to start with 1D functions, define curves in 2D planes and vary them in 3D space to create surfaces is given:

9.2.1 Function Catalog

A set of functions $Y(X)$ is suitably defined within the interval $0 < X < 1$, with end values at X,Y $= (0, 0)$ and $(1, 1)$, see Figure 53. We can imagine a multiplicity of algebraic and other explicit functions $Y(X)$ fulfilling the boundary requirement and, depending on their mathematical structure, allowing for the control of certain properties especially at the interval ends. Four parameters or less were chosen to describe end slopes (a, b) and two additional properties (e_G, f_G) depending on a function identifier G. The squares shown depict some algebraic curves where the additional parameters describe exponents in the local expansion (G=1), zero curvature without (G=2) or with (G=20) straight ends added, polynomials of fifth order (G=6, quintics) and with square root terms (G=7) allowing curvatures being specified at interval ends. Other numbers for G yield splines, simple Bezier parabolas, trigonometric and exponential functions. For some of them a, b, e_G and/or f_G do not have to be specified because of simplicity, like G=4 which yields just a straight line. The more recently introduced functions like G=20 give smooth connections as well as the limiting cases of curves with steps and corners. Implementation of these mathematically explicit relations to the computer code allows for using functions plus their first, second and third derivatives. It is obvious that this library of functions is modular and may be extended for special applications, the new functions fit into the system as long as they begin and end at (0, 0) and (1, 1), a and b - if needed - describe the slopes and two additional parameters are permitted.

$$Y = F_G(a, b, e_G, f_G, X)$$

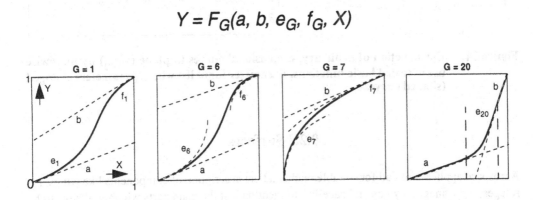

Figure 53   **Selection of 4 basic functions F_G in nondimensional unit interval**

9.2.2 Curves

The next step is the composition of curves by a piecewise scaled use of these functions. Figure

54 illustrates this for an arbitrary set of support points, with slopes prescribed in the supports and curvature or other desired property of each interval determining the choice of function identifiers G. The difference to using spline fits for the given supports is obvious: for the price of having to prescribe the function identifier and up to four parameters for each interval we have a strong control over the curve. The idea is to use this control for a more dedicated prescription of special aerodynamically relevant details of airframe geometry, hoping to minimize the number of optimization parameters as well as focusing on problem areas in CFD flow analysis code development. Numbers serving as names ("keys") distinguish between a number of needed curves, the example shows two different curves and their support points. Besides graphs a table of input numbers is depicted, illustrating the amount of data required for these curves. Nondimensional function slopes a, b are calculated from input dimensional slopes s_1 and s_2, as well as the additional parameters e_G, f_G are found by suitable transformation of e and f. A variation of only single parameters allows dramatic changes of portions of the curves, observing certain constraints and leaving the rest of the curve unchanged. This is the main objective of this approach, allowing strong control over specific shape variations during optimization and adaptation.

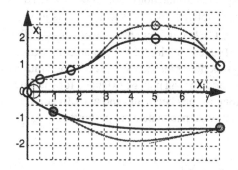

key	u	F(u)	s_1	G	s_2	e	f
1	0.0	0.0	0.	7	0.25	4.	0.
1	0.5	0.5	0.	4			
1	1.7	0.8	0.25	6	0.	0.	-0.2
1	5.0	2.0 / 2.5	0.	6	-0.8	-0.2	-0.2
1	7.5	1.0					
2	0.0	0.0	0.	7	-0.5	4.	0.
2	1.0	-0.7	-0.5	20	-0.2 / 0.2	5.	
2	7.5	-1.3					

Figure 54   **Construction of arbitrary, dimensional curves in plane (x_i, x_j) by piecewise use of scaled basic functions. Parameter input list with 2 parameters changed (shaded curves).**

9.2.3 Surfaces

Aerospace applications call for suitable mathematical description of components like wings, fuselages, empennages, pylons and nacelles, to mention just the main parts which will have to be studied by parameter variation. Three-view geometries of wings and bodies are defined by planforms, crown lines and some other basic curves, while sections or cross sections require additional parameters to place surfaces fitting within these planforms and crown lines. Figure 54 shows a surface element defined by suitable curves (generatrices) in planes of 3D space, it can be seen that the strong control which has been established for curve definition, is maintained here for surface slopes and curvature.

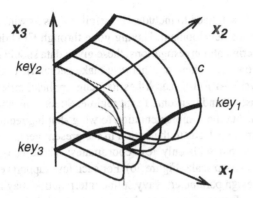

Figure 55   **Surface definition by cross sections c in plane (x_1, x_3) determined by generatrices (key$_i$), along x_2 and defined in planes (x_1, x_2) and (x_2, x_3).**

9.3 Aircraft and Aerospace Vehicle Components

So far the geometry definition tool is quite general and may be used easily for solid modelling of nearly any device if a mathematically exact description of the surface, with controlled gradients and curvatures, is intended. In aerodynamic applications we want to make use of knowledge bases from hydrodynamics and gasdynamics, i. e. classical airfoil and wing theory, as well as the classical results of slender body theory, transonic and supersonic area rule should determine the choice of functions and parameters. Surface quality should be described with the same accuracy as resulting from refined design methods outlined in the book chapter about the gasdynamic knowledge base. This is achieved by selecting suitable functions (G) when the 'key' curves are subdivided into intervals defined by support stations. Slope and curvature control avoids the known disadvantages of splines while at the same time the number of supports may be very low, if large portions may be modelled by one type of function.

In the process of making this generally described geometry tool to become dedicated geometry generator software for aerospace applications, a focusing on two main classes of surfaces has been found useful: There are classes of surfaces which are traditionally '*spanwise defined*' and others are '*axially defined*'. Lift-generating components like wings primarily belong to the first category while fuselages are usually of the second kind. With this distinction having led to several practical versions of geometry generators, it should not be considered too dogmatically: especially novel configuration concepts in the high Mach number flight regime are modelled without the above distinction as will be illustrated below, after describing the creating of conventional wings and fuselages.

9.3.1 Airfoils

In the case of wing design we will need to include 2D airfoil shapes as wing sections, with data usually resulting from previous development. Having gone through CFD design and analysis, sometimes also through experimental investigations, these given data should be dense sets of co-ordinates without the need to smooth them or otherwise make geometric changes which are not accompanied by flow analysis. Airfoil research has its main applications in high aspect ratio wing applications in the subsonic and transonic flight regime. Supersonic applications with low aspect ratio also need airfoils but their implementation to wing shaping requires mainly investigating the whole 3D problem. This leads to the option in the present geometry generator to provide again airfoil input data, but with only few coordinates: These can be used for spline interpolation in a suitably blown-up scale (Figure 56). For such few supports each point may take the role of an independent design parameter, wavy spline interpolation may be avoided if dislocations are small compared to distances to fixed points. Along one airfoil contour to be modified, portions of fixed contour with dense data distribution may be given while other portions may be controlled by only one or two isolated supports. This option was used in an early version of this geometry tool to optimize wing shapes in transonic flow [118].

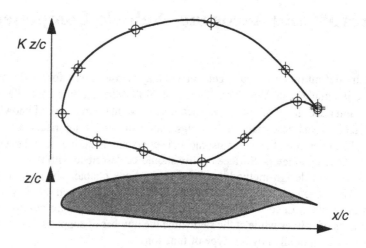

Figure 56   Spline fit obtained for airfoil in blown-up scale with few support points

Analytical sections and input for inverse design

Spline fits are well suited for redistribution of qualitatively acceptable dense data. The possible occurrence of contour wiggles has restricted their use in the geometry tool discussed here to the abovementioned option accepting external data, which is realistic for airfoils to be implemented in wing design. For a more independent approach we may ask for a more elegant analytical representation of wing sections, especially if these shapes still should be optimized. An important question arising is how many free parameters are needed for representation of arbitrary, typical wing sections, with the shape close enough to duplicate CFD or experimental results of aerodynamic performance with reasonable accuracy. Our successive refinement of airfoil generator subroutines using variously segmented curves as depicted in Figure 54 has shown that an amount of 10 to 25 parameters (numbers as listed in the table in Figure 54) may suffice for quite satisfactory representation of a given airfoil. The upper limit applies to transonic and laminar flow control airfoils with delicate curvature distribution as illustrated for a shock-free transonic airfoil, Figure 60 in chapter 7, where the influence of local curvature variations on the drag polar can be seen. The lower limit seems to apply for simpler yet practical subsonic airfoils and for most supersonic sections.

With a library of functions applied to provide parametric definition of airfoils, another application of this technique seems attractive: new inverse airfoil and wing design methods need input target pressure distributions for specified operation conditions and numerical results are found for airfoil and wing shapes. The status of these methods is reviewed in the next book chapter. Given the designer's experience in aerodynamics for selecting suitable pressure distributions, choice of a few basic functions and parameters may provide a dense set of data $c_p(x/c)$ just like geometry coordinates are prescribed, the amount of needed parameters for typical attractive pressure distributions about the same as for the direct airfoil modelling.

Variable camber sections

Lifting wings need mechanical control devices to vary their effective camber. Geometrical definition of simple hinged and deflected leading and trailing edges are defined by airfoil chordwise hinge locations and deflection angles. A more sophisticated mechanical flow control includes elastic surface components to ensure a certain surface smoothness across the hinge, such devices are called sealed slats and flaps (Figure 57). Spline portions or other analytical connection fits may suitably model any proposed mechanical device, an additional parameter is the chord portion needed for the elastic sealing.

Multicomponent airfoils

While sealed flaps and slats are suitable for supersonic wings, the much more complicated multicomponent high lift systems have been developed for current subsonic transport aircraft. In addition to angular deflection of slat and flap components, they require kinematic shifting devices housed within flap track fairings below the wing. For a mathematical and parameter-controlled description of slat and flap section geometries within the clean airfoil, the richness of our function catalog provides suitable shapes and track curves for a realistic modelling of these components in every phase of start and landing configurations. Figure 58 illustrates a multicomponent high lift system in 2D and 3D.

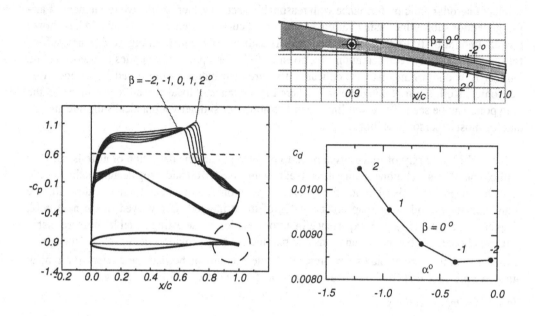

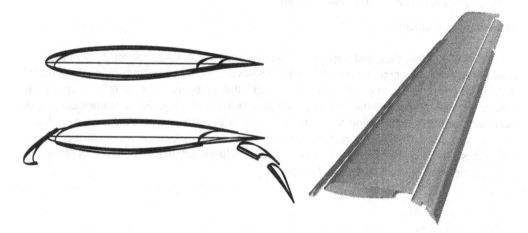

Figure 57   Variable camber sealed flap example. Flap deflection as a function of angle of attack variations, for constant lift. Airfoil in transonic flow, $M_\infty = 0.75$, Re = 40 Mill., $c_l = 0.7$, (MSES [93] analysis)

Figure 58   Wing sections for multicomponent high lift system, 3D swept wing with slat and flaps.

9.3.2 Spanwise defined components

Aircraft wings

Aerodynamic performance of aircraft mainly depends on the quality of its wing, design focuses therefore on optimizing this component. Using the present shape design method, we illustrate the amount of needed "key curves" along wing span which is inevitably needed to describe and vary the wing shape, Figure 59. The key numbers are just identification names: span of the wing y_0 in the wing coordinate system is a function of a first independent variable 0 < p < 1, the curve $y_0(p)$ is key 20. All following parameters are functions of this wing span: planform and twist axis (keys 21-23), dihedral (24) and actual 3D space span coordinate (25), section twist (26) and a spanwise section thickness distribution factor (27). Finally we select a suitably small number of support airfoils to form sections of this wing. Key 28 defines a blending function 0 < r < 1 which is used to define a mix between the given airfoils, say, at the root, along some main wing portions and at the tip. The graphics in Figure 59 shows how the basic airfoils, designed with subsonic or with supersonic leading edges, may be dominating across this wing. Practical designs may require a larger number of input airfoils and a careful tailoring of the section twist to arrive at optimum lift distribution, for a given planform.

Recent updates to the wing generation include a spanwise definition of the previously mentioned 10 - 25 airfoil parameters as additional key functions, replacing given support airfoils and the blending key 28. Because of an explicit description of each wing surface point without any interpolation and iteration, other than sectional data arrays describing the exact surface may easily be obtained very rapidly with analytical accuracy.

Wings with high lift systems are created using multicomponent airfoils either for unswept wings with simply their varied deflected 2D configurations as illustrated in Figure 57, or in the more practical case of swept components (Figure 58) rotation axes and flap tracks need to be described as lines and curves in 3D space. The clean airfoil configuration of the system is then changed observing the given 3D kinematics.

Other components with wing-type parameterization

Besides aircraft wings the tail and rudder fins as well as canard components are of course treatable with the same type of parameters and key functions. Highly swept and very short aspect ratio wing type components are the pylons for jet engines mounted to the aircraft wing; they need to be optimized in a flow critically passing between wing and engine. Generally any solid boundary condition to be optimized in flow with a substantial crossing velocity component is suitably defined as a spanwise defined component with a parameter set as illustrated for the wing.

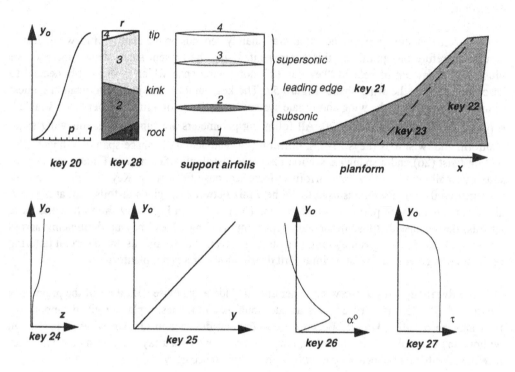

Figure 59   **Wing parameters and respective key numbers for section distribution, planform, an/dihedral, twist, thickness distribution factor and airfoil blending.**

9.3.3 Axially defined components

Fuselage bodies, nacelles, propulsion and tunnel geometries

This group of shapes is basically aligned with the main flow direction, the usual development is directed toward creating volume for payload, propulsion or, in internal aerodynamics (and hydrodynamics) the development of channel and pipe geometries. The parameters of cross sections are quite different to those of airfoils; the quality of their change along a main axis with constraints for given areas within the usually symmetrical contour is the design challenge. Fuselages are therefore described by another set of "keys" which is defined along the axis. This axis may be a curve in 3D space, with available gradients providing cross section planes normal to the axis. For simple straight axes in the cartesian x-direction key 40 defines axial stations just like key 20 defines spanwise x-stations. With the simplest cross section consisting of superelliptic quarters allowing a choice of the half axes or crown lines and body planform, plus the exponents (with the value of 2. for ellipses), 8 parameters (key 41 - 48) are given (Figure 60). Basic bodies are described easily this way, with either explicitly calculating the horizontal coordinate $y(x, z)$ for given vertical coordinate z, or the vertical upper and lower coordinate $z(x, y)$ for given points y within the planform, at each cross section station x = const.

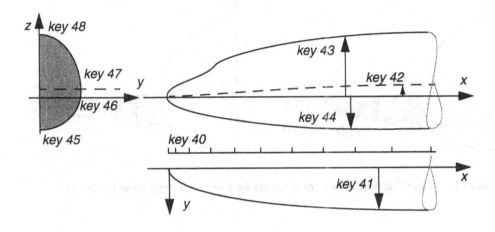

Figure 60   **Fuselage parameters and respective key numbers for cross section definition, planform and crown lines, superelliptic exponents.**

More complex bodies are defined by optional other shape definition subprograms with additional keys (49 - 59) needed for geometric details. These may be of various kind but of paramount interest is the aerodynamically optimized shape definition of wing-body junctures. In the following a simple projection technique is applied requiring only a suitable wing root geometry to be shifted toward the body, but more complex junctures require also body surface details to suitably meet the wing geometry.

9.3.4 Component intersections and junctures

The usual way to connect two components is to intersect the surfaces. Intersection curves are found only by numerical iteration for non-trivial examples. Most CAD systems perform such task if the data of different surfaces are supplied. Here we stress an analytical method to find not only the juncture curve but also ensure a smooth surface across the components avoiding corners which usually create unfavorable aerodynamic phenomena. Sketched in Figure 61, this can be applied generally to two components F_1 and F_2 with the condition that for the first component one coordinate (here the spanwise y) needs to be defined by an explicit function $y = F_1(x,z)$, while the other component F_2 may be given as a dataset for a number of surface points. Using a blending function for a portion of the spanwise coordinate, all surface points of F_2 within this spanwise interval may be moved toward the surface F_1 depending on the local value of the blending function. Figure 61 shows that this way the wing root (F_2) emanates from the body (F_1), wing root fillet geometry can be designed as part of the wing prior to this wrapping process. Several refinements to this simple projection technique have been implemented to the program.

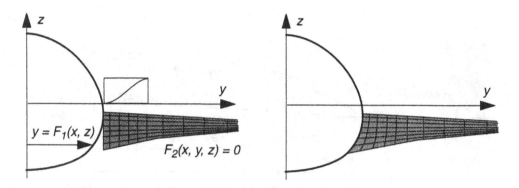

Figure 61   Combination of two components by a blended projection technique

9.3.5 Extensions to the fourth dimension

The outlined geometry generator based on this explicit mathematical function toolbox allows for creating models for nearly any aerospace-related configuration. The next step is to provide a whole series of shapes which result from a controlled variation of a parameter subset key_j, with the option to create infinitesimally small changes between neighboring surfaces. This requires the introduction of a "superparameter" t, its variation within a suitable interval Δt and a general variation function f(t). Variated parameters result then to

$$key_j(t) = key_j(0) + f(t) \ \Delta key_j \tag{71}$$

with Δkey defined by the chosen extreme deviation from the starting values. Obtaining a series of surfaces calls for suitable computergraphic animation technology [119]. There are three major applications of introducing the 4th dimension (t) to the presented geometry generator for the development of design concepts:

Numerical Optimization

The success of optimization performing variations of a set of parameters small enough to enable the designer to control and understand the evolutionary process toward improved performance, but large enough to most likely include a global optimum, depends on selecting the parameters by knowledge based criteria. Simple first applications include the calibration of surface modifications as experienced from shock-free transonic design [118], [120].

Adaptive Devices

A mechanical realization of numerical optimizing processes is the use of adaptive devices controlled by flow sensors [121]. Experiments are needed for the development and understanding of the dynamics of such processes, as they are already routine for adaptive wind tunnel walls. Adaptive configuration shape simulation by the geometry generator will require a series of shapes generated by selected functions equivalent to the mechanical model for elastic or pneumatic devices.

Unsteady configurations

Finally there is time, the natural role of the superparameter t. Configurations may vary with time, especially if there is aeroelastic coupling between structure and flow. Periodically varying shapes are generated to study the influence of moving boundary conditions on the flow. Modelling buffeting in the transonic regime is a wellknown goal, application of periodic geometries for a coupling of numerical structure analysis and CFD seems timely. Shape changes to model an adaptive helicopter rotor section with a sealed slat periodically drooped nose have been carried out and the results of unsteady Navier Stokes analysis suggest a concept for dynamic stall control [122].

9.4 Applications

Case studies for new generation supersonic transport aircraft have been carried out through the past years in research institutions and in the aircraft industry. Our present tool to shape such configurations needs to be tested by trying to model the basic features of various investigated geometries. Knowing that the fine-tuning of aerodynamic performance must be done by careful selection of wing sections, wing twist distribution and the use of sealed slats and flaps, with initial exercises we try to geometrically model some of the published configurations, generate CFD grids around them and use optimization strategies to determine the sensitivity of suitable geometry parameters. This is still a difficult task but tackling its solution greatly contributes to building up the knowledge base of high speed design.

9.4.1 Example: Generic High Speed Civil Transport Configuration

Figure 62 and Figure 63 illustrate data visualization of a generated configuration derived from a Boeing HSCT design case for Mach 2.4 [123]. The configuration consists of 10 components, engine pylons are not yet included. Wing and horizontal and vertical tail components are spanwise defined, fuselage and engines are axially defined components. The wing has a subsonic leading edge in the inner portion and a supersonic leading edge on the outer portion.

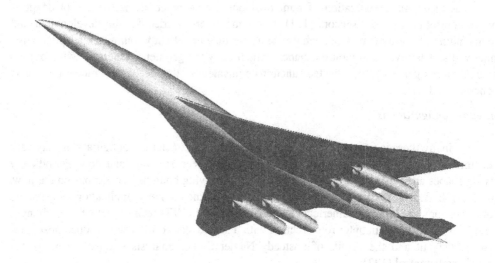

Figure 62   Generic HSCT configuration derived from Boeing Mach 2.4 case study: shaded graphic visualization of geometry modelling result

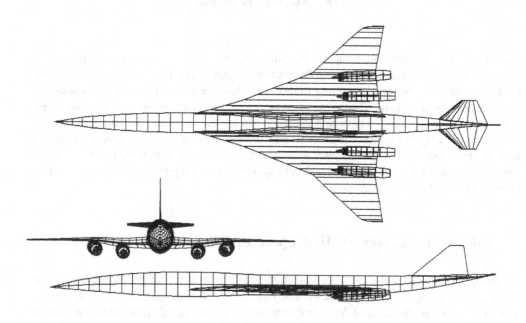

Figure 63   Generic HSCT configuration: Three-view wireframe model

For this study a minimum of support airfoils (Figure 58) is used to get a reasonable pressure distribution: a rounded leading edge section in most of the inner wing and a wedge-sharp section in the outer wing portion define the basic shape of the wing. Wing root fillet blending, the smooth transition between rounded and sharp leading edge and the tip geometry are effectively shaped by the previously illustrated wing keys, the fuselage here is a simple slender body requiring just the baseline body tool with elliptical cross sections.

Preprocessing input data for CFD requires providing a grid surrounding the configuration. For application of either structured or unstructured grids additional geometric shapes need to be provided. In the case of the generic HSCT with given supersonic flight Mach number the farfield boundary is chosen to engulf the expected bow shock wave (Figure 64a) and a cross sectional grid for both wing and body is generated, either as simple algebraic trajectories or using elliptic equations.

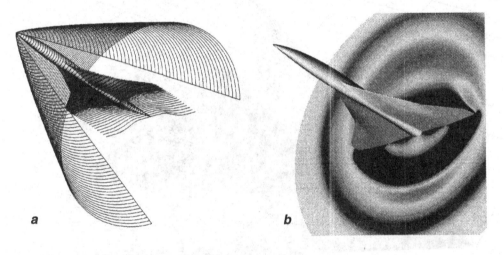

a   b

Figure 64   CFD grid boundaries, result for Euler analysis of HSCT wing-body in supersonic flow $M_\infty = 2.4$

Short runs using an the inviscid flow Euler option of a flow solver [124]were carried out on a coarse (33 x 81 x 330) grid, here only to get an idea about the needed wing section and twist modifications for acceptable pressure distributions. Visualization of the results in various grid surfaces is needed, like pressure distributions in cross section planes as shown in Figure 64b.

Coupling with visualization tools

Postprocessing of CFD results with a powerful graphic system [117] shows detailed display of flow variables distribution along the configuration and in the flow field. Selected cross section pressure checks allow for an assessment of chosen airfoils and twist distributions before refined grids and longer Euler or Navier-Stokes runs are executed. Though areas of necessary local grid refinement are spotted, some basic information about needed airfoil changes is already provided by such short runs; the refinement of geometry and CFD analysis may begin.

Visualization of the shock waves system emanating from the body tip and the wing is shown in Figure 65. A new visualization technique [125] allows for analyzing shock waves found by CFD analysis in 3D space: their quality near the aircraft, as shown, or with refined CFD analysis in larger distances to investigate sonic boom propagation, may be a useful help to assess this environmentally important aspect of supersonic transport. The figure shows a cut-off domain of the shock surfaces: A shock strength threshold allows analysis of local sonic boom quantities.

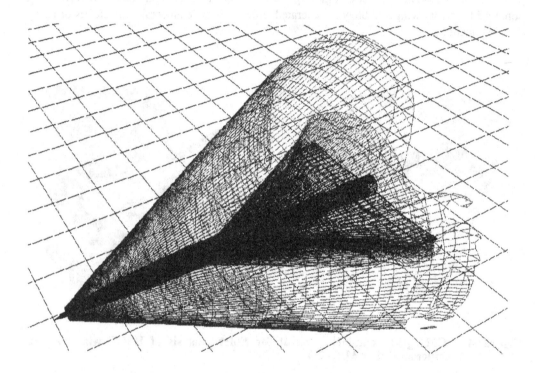

Figure 65   **Visualization of CFD results: Shock system emanating from body tip and wing root, cut-off at threshold for selected sonic boom strength.**

9.4.2 Software development for CFD and CAD preprocessing

The above case study, to be further used for a refined analysis and design parameter identification, is just a modest example compared to the needed studies in the development of industrial software suitable for the envisioned multidisciplinary quality as outlined in the first book chapters. With the proven flexibility of the ideas developed here, a larger scale software development at industry, aerospace research establishment and university institutes has picked up the basic elements and merged them in their applied software systems:

At Daimler Benz Airbus aircraft industry a software system was developed as a pre-processor for multiblck structured grid generation around complete aircraft [126]. The fast algebraic definition of arbitrary surface metrics is ideal for application in multiblock/multigrid CFD analysis; many case studies exploiting the richness of possible configuration topologies with this approach have been carried out. Refinements in the algebraic grid generation tools for optimum multiblock grid spacing have been implemented and complex transonic transport aircraft configurations with suitable grid blocks have been generated and used for CFD analysis.

Parametric studies of supersonic transport aircraft wings have been performed at DLR German Aerospace Research Establishment [127]. Sensitivity studies are carried out and a multi-point optimization design method is being worked on using some of our basic functions and curves. Results are obtained for conventional and new configurations, with optima found for relative positioning of the different components.

The emergence of new programming languages and faster and more powerful graphic workstations with larger storage capacity gave rise to the development of a new and completely interactive version of this geometry generator [128]. Definition of a complex case study for a combined theoretical, numerical and experimental investigation sets various tasks for the new system to serve as a preprocessor for different CAD systems. These systems are needed for wind tunnel model construction (CATIA) and for unstructured CFD grid generation (ICEM). The latter is needed for performing Euler and Navier Stokes analysis with a new analysis code using unstructured grids [129]. Figure 66 illustrates a result of this analysis.

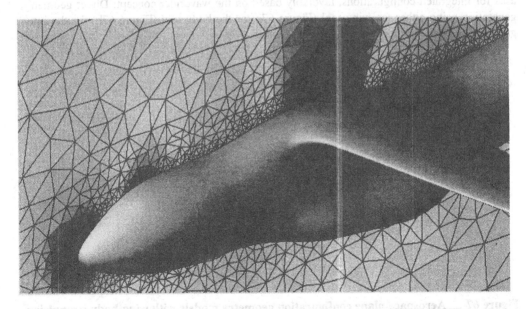

Figure 66   **Generic high wing transport aircraft geometry model for Euler CFD analysis with unstructured grid code; grid and pressure isofringes in center plane**

9.4.3 Novel Configurations

The development of conventional high speed transport configurations like the generic HSCT configuration may still face crucial technology problems resulting in reduced chances to operate economically, as it is critically reviewed in various chapters in this book. New concepts, on the other hand, are emerging, but they must be studied in great detail using reliable theoretical, numerical and experimental analysis tools before any project can be laid out for development of a first aircraft.

All-body and all-wing configurations

From the viewpoint of using the presented geometry generator for support of such new concepts, it seems that promising new configurations can be generated if the two types of shapes, axially and spanwise defined components, are not any longer restricted to their traditional roles of representing fuselage and wing, respectively. There is rather an attractive alternative emerging by either type of component taking over both functions:

All-body aircraft as well as all-wing configurations are limiting cases where either wing or fuselage is vanishing and the remaining component taking over both functions which are providing volume as well as lift. Waveriders as illustrated in the previous book chapter are fully integrated configurations; guided by the outlined knowledge base of inverse design it is now relatively easy to create arbitrary direct design cases with waverider characteristics [115], there is just no "on-design" condition flow field coming with the design geometry. Suitable choice of geometry parameters to simulate inverse design cases but allowing a 4D optimization extension as outlined above most likely will lead to further improvements. Generic hypersonics asks for integrated configurations, favorably based on the waverider concept: Direct geometry generation using either the wing tool (Figure 67a) or the body tool (Figure 67b) for the integrated wing body components can solve this task.

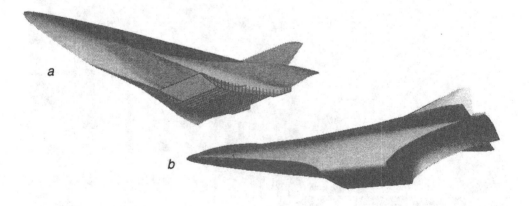

Figure 67   **Aerospace plane configuration geometry models with wing-body-propulsion integration.**

All-wing or Flying Wing aircraft has several advantages reviewed in other chapters of this book; here its attractivity for both aerodynamics and structures in high speed flight just means that we may focus on case studies to model a variety of such Flying Wings as input for detailed analysis in a multidisciplinary approach.

Oblique Flying Wing

A shape with a relatively simple geometry at first sight is the Oblique Flying Wing (OFW), an ultimate example of adaptive geometry by adjusting the yaw angle of the whole configuration (exept the engines and control surfaces) to the varying flight Mach numbers (Figure 68). After several conclusions about the attractivity of this concept in this book, the two final chapters are entirely devoted to the OFW. In the last chapter some studies are presented using our geometry software for OFW definition. Challenging tasks for systematic geometry parameter fine tuning emerge from the obtained results. Ongoing work will profit from a combination of this geometry generator with optimization tools as outlined in the following chapters.

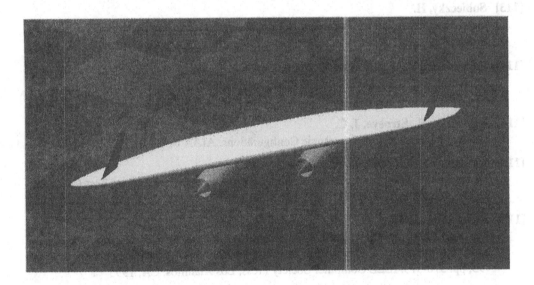

Figure 68   **Oblique Flying Wing model, with control surfaces and propulsion adjusted to the flight direction.**

9.5 Conclusions

Software for generic aerodynamic configurations has been developed to support the design requirements in the high speed regime. Based on simple, explicit algebra a set of flexible model functions is used for curve and surface design which is tailored to create realistic airplanes or their components with various surface grid metrics. The explicit and non-iterative calculation of surface data sets make this tool extremely rapid and this way suitable for generating whole series of configurations in optimization cycles. The designer has control over parameter variations and builds up a knowledge base about the role of these parameters influencing flow quality and the aerodynamic performance coefficients. Gasdynamic relations and other model functions allow for the gradual development of our design experience if generic configurations are used as boundary conditions for numerical analysis with mature CFD codes. Experimental investigations are supported by CAD data which are delivered from the same geometry inputs as used for preprocessing numerical simulation. With efficient geometry tools available to the designer, the development of interactive design systems for not only aerodynamic but multidisciplinary optimization gets additional momentum.

9.6 References

[113] **Sobieczky, H.**
       Geometry Generation for Transonic Design. Recent Advances in Numerical Methods in Fluids, Vol. 4, Ed. W. G. Habashi, Swansea: Pineridge Press, pp. 163 - 182, 1985

[114] **Pagendarm, H. G., Laurien, E., Sobieczky, H.**
       Interactive Geometry Definition and Grid Generation for Applied Aerodynamics. AIAA 88-2415CP, 1988

[115] **Sobieczky, H., Stroeve, J. C.**
       Generic Supersonic and Hypersonic Configurations. AIAA 91-3301CP, 1991

[116] **Sobieczky, H.**
       DLR-F5: Test Wing for CFD and Applied Aerodynamics, Case B-5 in: Test Cases for CFD Evaluation. AGARD FDP AR-303, 1994

[117] **Pagendarm, H. G.**
       Unsteady Phenomena, Hypersonic Flows and Co-operative Flow Visualization in Aerospace Research. In: G.M. Nielson, D. Bergeron, (Editors), Proceedings Visualization '93, pp. 370-373, IEEE Computer Society Press, Los Alamitos, CA, 1993

[118] **Cosentino, G. B., Holst, T. L.**
       Numerical Optimization Design of Advanced Transonic Wing Configurations. Journal of Aircraft, Vol. 23, pp. 192-199, 1986

[119] **Hannemann, M., Sobieczky, H.**
       Visualization of High Speed Aerodynamic Configuration Design. In: G. M. Nielsen, D. Silver, Proc. Visualization '95, pp. 355 - 358, IEEE Computer Society Press, Los Alamitos, CA, 1995

[120] **Zhu, Y., Sobieczky, H.**
       Numerical Optimization Method for Transonic Wing Design. Proc. 6th Asian Congress of Fluid Mechanics, Singapore, 1995

[121] **Sobieczky, H., Seebass, A. R.**
Adaptive Airfoils and Wings for Efficient Transonic Flight. ICAS paper 80-11.2, 1980

[122] **Geissler, W., Sobieczky, H.**
Unsteady Flow Control on Rotor Airfoils. AIAA 95-1890, 1995

[123] **Kulfan, R.**
High Speed Civil Transport Opportunities, Challenges and Technology Needs. Lecture at Taiwan IAA 34th National Conference, 1992

[124] **Kroll, N., Radespiel, R.**
An Improved Flux Vector Split Discretization Scheme for Viscous Flows. DLR-FB 93-53, 1993

[125] **Pagendarm, H. G., Choudhry, S. I.**
Visualization of Hypersonic Flows - Exploring the Opportunities to Extend AVS, 4th Eurographics Workshop on Visualization in Scientific Computing, 1993

[126] **Becker, K., Rill, S.**
Multiblock Mesh Generation of Complete Aircraft Configurations. In: N. P. Weatherill, M. J. Marchant, D. A. King (Editors), Multiblock Grid Generation Results of the EC/ BRITE-EURAM EUROMESH, 1990-92. Notes on Numerical Fluid Mechanics, Vol. 44, pp. 130 - 138, Braunschweig; Wiesbaden: Vieweg, 1993

[127] **Orlowsky, M., Herrmann, U.**
Aerodynamic Optimization of Supersonic Transport Configurations. Proc. 20th International Council of the Aeronautical Sciences Congress, ICAS 96.4.4.3, 1996

[128] **Trapp, J., Zores, R., Gerhold, Th., Sobieczky, H.**
Geometrische Werkzeuge zur Auslegung Adaptiver Aerodynamischer Komponenten. Proc. DGLR Jahrestagung '96, Dresden, 1996

[129] **Gerhold, Th., Friedrich, O., Galle, M.**
Calculation of Complex Three-Dimensional Configurations Employing the DLR-Tau-Code. AIAA 97-0167, 1997

CHAPTER 10

AERODYNAMIC SHAPE INVERSE DESIGN METHODS

G.S. Dulikravich
The Pennsylvania State University, University Park, PA, USA

10.1 Introduction

Aerodynamic problems are defined by the governing partial differential or integral equations, shapes and sizes of the flow domains, boundary and initial conditions, fluid properties, and by internal sources and external inputs of mass, momentum and energy. In the case of an analysis (direct problem) we are asked to predict the details of a flow-field if the shape(s) and size(s) of the object(s) are given. In the case of a design (inverse or indirect problem) we are asked to determine the shape(s) and size(s) of the aerodynamic configuration(s) that will satisfy the governing flow-field equation(s) subject to specified surface pressure or velocity boundary conditions and certain geometric constraints [130]-[138]. The entire design technology is driven by the increased industrial demand for reduction of the design cycle time and minimization of the need for the costly a posteriori design modifications.

Aerodynamic inverse design methodologies can be categorized as belonging to surface flow design and flow-field design. Surface flow design is based on specifying pressure, Mach number, etc. on the surface of the object, then finding the shape of the object that will generate these surface conditions. Flow-field design enforces certain global flow-field features (shock-free conditions, minimal entropy generation, etc.) at every point of the flow-field by determining the shape that will satisfy these constraints. An arbitrary distribution of the surface flow parameters or an arbitrary field distribution of the flow parameters could result in aerodynamic shapes that either cross over ("fish tail" shapes) or never meet ("open trailing edge" shapes). These problems can be avoided by appropriately constraining the surface distribution of the flow parameters [139].

It should be pointed out that inverse methods for aerodynamic shape design are capable of creating only point-designs, that is, the resulting shapes will have the desired aerodynamic characteristics only at the design conditions. If the angle of attack, free stream Mach number, etc. in actual flight situations are different from the values used in the design, the aerodynamic performance will deteriorate sometimes quite dramatically. For example, when designing transonic shock-free shapes with a surface flow design method, the resulting configuration could have a mildly concave part of its surface locally covered by the supersonic flow indicating the existence of a "hanging shock" or a "loose-foot" shock [139] even at the design conditions. At off-design, the hanging shock attaches itself to the aerodynamic surface causing a boundary layer separation. Consequently, it is more appropriate to design shapes that have a weak family of shocks [140] since such designs have been found not to increase the shock wave strengths appreciably at off-design conditions.

In this chapter, we will focus on briefly explaining only these aerodynamic shape inverse design concepts that are applicable to the design of three-dimensional (3-D) high speed configurations. An attempt will be made to focus on the techniques that have been found to be cost effective, reliable, easy to comprehend and implement, transportable to different computers, and accurate.

10.2 Surface Flow Data Specification

Once the global aerodynamic parameters (inlet and exit pressures, temperatures, and flow angles) have been specified, the next objective is to determine the best way to distribute aerodynamic quantities on the yet unknown configuration. Since the inverse shape design is based on the specified ("desired" or "target") surface pressure distribution, the common dilemma is the choice of the "best" surface target pressure. Specifically, it would be desirable to determine the best pressure distribution on the surface of the yet unknown configuration so that the aerodynamic efficiency is maximized by minimizing all possible contributions to the entropy generation in the entire flow-field. From the specified surface distribution of flow-field parameters it is possible to discern only certain aspects of the boundary layer. It is well known that the separated boundary layer significantly increases flow-field vorticity and, consequently, the viscous dissipation function, entropy generation and aerodynamic drag. To minimize these effects the desired surface pressure distribution can first be checked for possible flow separation before it is further used in the aerodynamic shape inverse design. A very fast method for detecting flow separation has been recently proposed [141]. It is based on the fact that the rate of change of flow kinetic energy reaches its minimum at the separation point. The kinetic energy can be calculated from the surface pressure distribution by assuming that pressure does not change across a boundary layer. The flow separation detection code is short and very simple since it involves algebraic and analytic expressions only. Thus, the surface pressure distribution, either specified by the designer or obtained while using an optimization process, can be quickly checked for possible flow separations (Figure 69) before it is actually enforced.

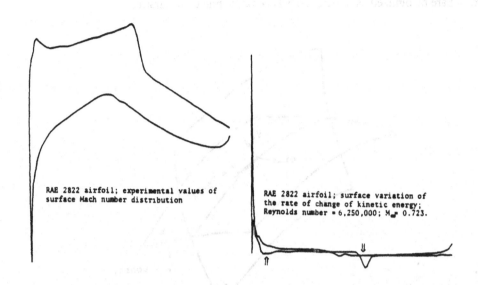

RAE 2822 airfoil; experimental values of
surface Mach number distribution

RAE 2822 airfoil; surface variation of
the rate of change of kinetic energy;
Reynolds number = 6,250,000; M_∞ 0.723.

Figure 69   **Detection of flow separation locations from the specified or measured surface
pressure distribution [141]. Arrows point at experimentally found locations
of the flow separation points. Our code predicts separation to occur at the
minimum of the curve representing the local rate of change of surface flow
kinetic energy.**

This a priori checking can be automatically followed by the minor modifications of the
local surface pressure distribution with the objective of moving the flow separation points further downstream. The updated surface pressure distribution can be re-checked for the locations
of the flow separation points. This checking/modification procedure can be repeated until the
separation points cannot move downstream any more. This procedure is extremely fast since it
is accomplished without a single call to the flow-field analysis code.

10.3 Stream-Function-Coordinate (SFC) Concept

One of the fastest known inverse design techniques is based on a stream function formulation
[142]. The inviscid, compressible, steady flow around a given 3-D configuration can be predicted
by solving for two stream functions, $\Psi(x,y,z)$ and $\Lambda(x,y,z)$ [142]-[146] (Figure 70). For the purpose of inverse shape, design this formulation can be inverted. That is, two quasi-linear second
order coupled partial differential equations of the mixed elliptic-hyperbolic type can be derived.
These two equations treat the x-coordinate and the values of $\Psi(x,y,z)$ and $\Lambda(x,y,z)$ or their derivatives as known quantities on the, as yet, unknown solid surface of the 3-D configuration, while
treating values of $y = y(x,\Psi,\Lambda)$ and $z = z(x,\Psi,\Lambda)$ for every 3-D streamline as the unknowns. Thus,
the result of a numerical integration of this inverted system are the y-and-z coordinates of the 3-

D streamlines. Those streamlines that correspond to the specified surface values of $\Psi(x,y,z)$ and $\Lambda(x,y,z)$ are recognized as the desired 3-D aerodynamic configuration.

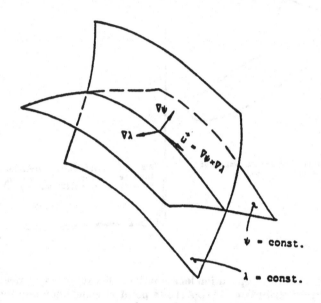

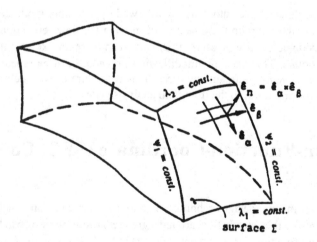

$$\dot{m} = \int_{\Sigma} \rho \vec{q} \cdot d\vec{A} = \int_{\Sigma} \nabla \psi \times \nabla \lambda \cdot d\vec{A} = \int_{\Sigma} |J| \, d\alpha \, d\beta = \int_{\Sigma} d\psi \, d\lambda = |(\psi_2 - \psi_1)(\lambda_2 - \lambda_1)|$$

Figure 70   **Geometrical interpretation of mass flow rate through a stream tube bounded by two pairs of stream surfaces [144].**

A computer code that implements this technique converges very fast because it implicitly satisfies mass at every iteration step thus avoiding the need for integrating the mass conservation equation. Moreover, such a code can be executed in an analysis mode when Dirichlet boundary conditions for $\Psi(x,y,z)$ and $\Lambda(x,y,z)$ are specified at every surface point, or in its inverse design mode when Neumann boundary conditions for $\Psi(x,y,z)$ and $\Lambda(x,y,z)$ are specified at every yet unknown surface point. Despite its remarkable speed of execution, robustness and the fact that the entire field of 3-D streamlines is obtained as a by-product of the computation, the SFC concept has its serious disadvantages. This inverse design method requires development of an entirely new code for the solution of the two Stream-Function-as-a-Coordinate (SFC) equations. The method is not applicable to viscous flow models, it suffers from the difficulties of the geometric multivaluedness of the stream functions, and is analytically singular at all of the points where the Jacobian of transformation $(\Psi,\Lambda,0)/(x,yz)$ become zero [142]. This occurs at every point where the x-component of the local velocity vector is zero which can happen at a number of points whose locations we do not know in advance. Consequently, the SFC method is recommended only for the inverse design of smooth 3-D configurations where preferably there are no stagnation points (3-D duct [146]) or where we are willing to neglect the designed shape in the vicinity of the leading and trailing edges.

10.4 Elastic Surface Motion Concept

There are many reliable and relatively fast 3-D flow-field analysis codes in existence. It would be highly economical to utilize these analysis codes in an inverse shape design process. This would require development of a short and simple design code that can utilize any of the existing flow-field analysis codes as a large, exchangeable subroutine. One very simple method for developing such a design code is based on utilizing an elastic membrane as a mathematical model. This concept treating those parts of the surface of the aerodynamic flight vehicle that are desired to have a specified surface pressure distribution as a membrane loaded with unsteady point-forces, ΔC_p, that are proportional to the local difference between the computed and the specified coefficient of surface pressure. Under such an unsteady load the membrane will iteratively deform until it assumes a steady position that experiences zero forcing function at each of the membrane points. Since this is a general non-physical concept for modeling the unsteady damped motion of the 3-D aerodynamic surface, any analytical expression governing damped motion of a continuous surface will suffice. Garabedian and McFadden [147] suggested a simple second-order linear partial differential equation as such a model where ΔC_p is proportional to the local surface slopes and curvatures.

$$\Delta n + \beta_1 \cdot \frac{\partial}{\partial x}\Delta n + \beta_2 \cdot \frac{\partial}{\partial y}\Delta n + \beta_3 \cdot \frac{\partial^2}{\partial x^2}\Delta n + \beta_4 \cdot \frac{\partial^2}{\partial y^2}\Delta n = \beta_5 \cdot \Delta C_p \qquad (72)$$

Here, the unknowns are the local normal surface displacements, Δn. If the surface grid point in question is on the upper surface, $\beta_5 = 1.0$, if on the lower surface, $\beta_5 = -1.0$. The remaining coefficients β_1 through β_4 are user-specified quantities that can accelerate the approach to a steady state. This partial differential equation can be discretized using finite difference representations for the partial derivatives. This leads to one penta-diagonal system or a sequence of two three-diagonal systems of algebraic equations that can readily be solved for the unknown normal surface modifications, Δn_i. This simple technique was successfully used to design isolated 3-D transonic supercritical wings [148] with a 3-D full potential code (Figure 71) as the flow analysis module. A simplified version of this concept (with Δz instead of Δn) was used to design engine nacelles and wing-body configurations [149]. The initial guess for the shape of a 3-D body does not have to be close to the final configuration for this method to work. Although requiring only 30-100 calls to a flow-field analysis code when using a panel code or a full potential equation code, the stiffness of the iterative matrix in the present formulation of the elastic membrane concept increases rapidly with the non-linearity of the flow-field analysis code used. This means that when using Euler or Navier-Stokes flow-field analysis code we will need between two and three orders of magnitude more calls to the flow-field analysis code than when using a simple linear panel code. For example, a two-dimensional airfoil shape inverse design with this method utilizing a Navier-Stokes flow analysis code may require over ten thousand calls to the Navier-Stokes code [150].

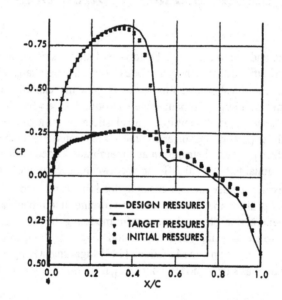

Figure 71   Comparison of baseline, target, and design surface pressures for a transonic purposely shocked airfoil using elastic surface motion concept [148].

10.5 Indirect Surface Transpiration Concept

This is one of the most common and oldest methods for aerodynamic shape inverse design. Like any iterative technique, it requires an initial guess for the aerodynamic shape. Then, using an inviscid flow-field analysis code with desired surface tangential velocity components enforced will result in non-zero values of the surface normal velocity component. The objective is then to find a configuration that has zero velocity components normal to the final body surface. The computed normal velocity components, v_n, are therefore used to modify the shape of the initially guessed configuration using a surface transpiration analogy. The new surface shape is predicted by treating the old surface as porous, hence fictitiously injecting the mass ($\rho \, v_n$) normal to the original surface so that the new surface becomes an updated stream surface. The local surface displacements, Δn, can be obtained from mass conservation equations for the quasi two-dimensional sections (stream tubes) of the flow-field bounded by the two consecutive cross sections of the body surface, the original surface shape, and the updated surface shape displaced locally by Δn [149]-[153]. Starting from a stagnation line where $\Delta n_{i-1,j} = 0.0$ and $\Delta n_{i-1,j+1} = 0.0$, separate updating of the pressure surface and the suction surface can be readily performed by solving for $\Delta n_{i,j}$ and $\Delta n_{i,j+1}$ from a bi-diagonal system. With the classical transpiration concept, normal surface velocities can be computed using any potential flow solver including a highly economical surface panel flow [152], [153] analysis code Figure 72. The indirect surface transpiration method works quite satisfactory in conjunction with Euler and even Navier-Stokes equations barring any shock waves or flow separation. A drawback of this approach is that during the repetitive surface updating using this method, the updated surfaces develop a progressively increasing degree of oscillation. This can be eliminated by periodically smoothing the updated surfaces with a least-squares surface fitting algorithm.

10.6 Direct Surface Transpiration Concept

This is an equally simple method of utilizing the transpiration concept which is especially applicable to viscous flow solvers, but the manner of treating boundary conditions on surfaces is different. For example, in the flow analysis with Navier-Stokes equations, the no-slip condition is imposed at the solid surface by enforcing zero values of all three components of the contravariant velocity vector defined as $U = D\xi/Dt$, $V = D\eta/Dt$, $W = D\zeta/Dt$. The ξ, η, ζ curvilinear, non-orthogonal coordinate system follows the structured computational grid lines. For example, let η-grid lines (thus V contravariant velocity component) emanate from the surface of the 3-D object under design consideration. In this inverse shape design concept, the surface pressure distribution, which is obtained from the specified pressure coefficient distribution, is enforced iteratively together with $U = W = 0$. This can be done readily in any existing Navier-Stokes flow analysis code [154]. It will result in the contravariant velocity vector component, V, becoming non-zero at the surface. Hence, the surface will have to move with iterations or time steps until the convergence is reached, that is, until $V = 0$ is satisfied on the final surface configuration. This will require that the computational grid be regenerated with each update of the surface. Thus, the aerodynamic parameters need to be transfered between old and new grid points by an accurate 3-D interpolation.

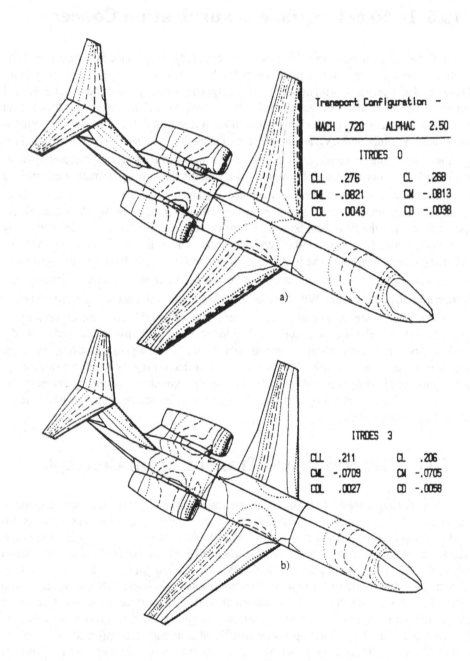

Figure 72   **Inverse design of an entire business jet configuration using a high order surface panel code and indirect surface transpiration concept can be performed on a personal computer in less than one hour: a) before, and b) after three optimization cycles enforcing the desired surface pressures. Notice improvements in aerodynamic coefficients [153].**

This design method will provide for a time-accurate motion of the solid boundaries if the code is executed in a time-accurate mode. For example, it is possible with this method to design a flexible 3-D shape that maintains an essentially steady surface pressure distribution in an otherwise unsteady flow. Specifically, it is possible to determine the correct instantaneous values of local swelling and contraction of a "smart" material coating on the surface thus creating the smart or continuously-adaptable aerodynamic shape design.

10.7 Characteristic Boundary Condition Concept

The correct number and types of boundary conditions for a system of partial differential equations can be determined by analyzing eigenvalues and the corresponding eigenvectors of the system in each coordinate direction separately. This approach to boundary condition treatment is called characteristic boundary conditions [155]-[157] since it suggests one-dimensional application of certain Riemann invariants at the boundaries that can be either solid or open boundaries. For example, in a 3-D duct flow, if the flow is locally subsonic at the exit, we will be able to compute all flow variables at the exit based on the information from the interior points except for one variable that we will have to specify at the exit. If the same characteristic boundary condition procedure is applied in the direction normal to the solid wall and if the desired pressure distribution is specified on the wall, this method of iteratively enforcing the boundary conditions at the wall will result in non-zero normal velocities at the wall which can be used to update the wall shape. The general concept follows.

The Euler equations for 3-D compressible unsteady flows expressed in non-conservative form and cast in a boundary-conforming, non-orthogonal, curvilinear (ξ, η, ζ) coordinate system can be transformed into

$$\frac{\partial Q}{\partial \tau} + B \cdot \frac{\partial Q}{\partial \eta} + D = 0 \tag{73}$$

where $Q = (\rho\ p\ u\ v\ w)$ is the transposed vector of the non-conservative primitive variables. Eigenvalues of B are

$$V - a \cdot \sqrt{\left(\eta_x^2 + \eta_y^2 + \eta_z^2\right)}\ V, V, V, V + a \cdot \sqrt{\left(\eta_x^2 + \eta_y^2 + \eta_z^2\right)} \tag{74}$$

where $V = \eta_t + \eta_x \cdot u + \eta_y \cdot v + \eta_z \cdot w$ and the local speed of sound is defined as $a = (\gamma\, p\, /\, \rho)^{1/2}$. If the η-grid lines are emanating from the 3-D aerodynamic configuration, we can have several situations. If $0 < V < a \sqrt{\left(\eta_x^2 + \eta_y^2 + \eta_z^2\right)}$, one eigenvalue is negative requiring a pressure boundary condition to be specified at that surface point.

Similarly, if-a $\sqrt{\left(\eta_x^2 + \eta_y^2 + \eta_z^2\right)} < V < 0$, four eigenvalues will be negative requiring pressure, velocity ratio $u/(u^2 + v^2 + w^2)^{1/2}$, total pressure and total temperature to be specified at that surface point. This method has been shown to converge quickly for transonic two-dimensional airfoil shape design when using compressible flow Euler equations [156],[157]. The method might be applicable to the inverse design of arbitrary 3-D configurations [155] although no such attempts have been reported yet. If the Euler code is executed in a time-accurate mode, the specified unsteady solid wall characteristic boundary conditions will provide for a time-accurate motion of the solid boundaries which is highly attractive for the design of "smart" aerodynamic configurations. This concept is not directly applicable to viscous flow codes since velocity components at the solid wall are zero.

10.8 Integro-Differential Equation Concept

An attractive property of integral equations is that the influence of the boundary conditions is transmitted throughout the flow-field instantaneously in the case of a linear flow problem. Even for non-linear flow problems the influence of the boundary conditions is transmitted throughout the flow-field extremely quickly as compared to the partial differential equation models where the finite difference or finite element discretization allows the influence of the boundary conditions to be transmitted at most one grid cell per during each iteration. A very fast and versatile 3-D aerodynamic shape inverse design algorithm was developed and is widely utilized in several countries [158]-[161]. It can accept any available 3-D flow-field analysis code as a large subroutine to analyze the flow around the intermediate 3-D configurations. The configurations are updated using a fast integro-differential formulation where a velocity potential perturbation $\phi(x,y,z)$ around an initial 3-D configuration $z_{+/-}(x,y)$ can be obtained from, for example, a Navier-Stokes code [161]. Here, the subscripts +/- refer to the upper and lower surfaces of the flight vehicle. Transonic 3-D small perturbation equation is

$$\frac{\partial^2}{\partial x^2}\Delta\phi + \frac{\partial^2}{\partial y^2}\Delta\phi + \frac{\partial^2}{\partial z^2}\Delta\phi = \frac{1}{V_\infty} \cdot \frac{\partial}{\partial x} \cdot \left(\frac{1}{2} \cdot \left(\frac{\partial\phi}{\partial x} + \frac{\partial}{\partial x}\Delta\phi \right)^2 - \frac{1}{2} \cdot \left(\frac{\partial\phi}{\partial x} \right)^2 \right) = \frac{1}{V_\infty} \cdot \frac{\partial\Gamma}{\partial x} \qquad (75)$$

Here, differentially small potential perturbation is $\Delta\phi(x,y,z)$ and x,y,z coordinates have been scaled via Prandtl-Glauert transformation, V_{∞} is the free stream magnitude, while

$$\frac{\partial\phi}{\partial x}(x, y, +/-0) = -\frac{V_\infty}{2} \cdot C_p(x, y) \qquad (76)$$

$$\frac{\partial}{\partial x}\Delta\phi(x, y, +/-0) = -\frac{(1 + \kappa) \cdot M \cdot a_\infty^2}{2 \cdot \left(1 - M \cdot a_\infty^2\right)} \cdot \left(C_{p+/-}^{spec} - C_{p+/-}^{calc}\right) \tag{77}$$

Here, M is the local Mach number and a_{oo} is the free stream speed of sound. The flow tangency condition is then

$$\frac{\partial}{\partial z}\Delta\phi(x, y, +/-0) = V_\infty \cdot \frac{\partial}{\partial x}\Delta z_{+/-}(x, y) \tag{78}$$

Since $\frac{\partial}{\partial z}\Delta\phi(x, y, +/-0)$ can be obtained from equation (75), the 3-D geometry is readily updated from

$$\Delta z_{+/-}(x, y) = \frac{1}{2} \cdot \int \frac{\partial}{\partial x}[\Delta z_+(x, y) + \Delta z_-(x, y)]dx \pm \frac{1}{2} \cdot \int \frac{\partial}{\partial x}[\Delta z_+(x, y) - \Delta z_-(x, y)]dx \tag{79}$$

Since equation (75) is linear, it can be reformulated using Green's theorem as an integro-differential equation. The Γ term on the right hand side of equation (75) would require volume integration which can be avoided if Γ is prescribed as smoothly decreasing away from the 3-D flight vehicle surface where it is known. Then, the problem can be very efficiently solved using the 3-D boundary element method. This inverse shape design concept has been successfully applied to a variety of planar wings [158]-[160] and wing-body configurations including the H-II Orbiting Plane with winglets [161], [162] where 3-D flow-field analysis codes were of the full potential, Euler and Navier-Stokes type. The method typically requires 10-30 flow analysis runs with an arbitrary flow solver and as many solutions of the linearized integro-differential equation.

Figure 73   **Winglets on a Japanese space plane were successfully redesigned using the integro-differential equation approach and a Navier-Stokes flow-field analysis code [161].**

10.9 Conclusions

Several prominent and proven methods that are applicable to inverse design of 3-D aerodynamic shapes have been briefly surveyed. The design computer codes based on these methods can be readily developed by modifying solid boundary condition subroutines in most of the existing flow-field analysis codes. Thus, all of the design methods surveyed are computationally economical since they require typically only a few dozen calls to the 3-D flow-field analysis code. Although the inverse shape design methods generate only point-designs, it was pointed out that at least two of the methods are conceptually capable of inverse shape design for unsteady flow conditions.

10.10 References

[130] **Dulikravich, G. S. (editor)**
Proceedings of the 1st International Conference on Inverse Design Concepts in Engineering Sciences (ICIDES-I), University of Texas, Dept. of Aero. Eng. & Eng. Mech., Austin, TX, October 17-18, 1984.

[131] **Dulikravich, G. S. (editor)**
Proceedings of the 2nd International Conference on Inverse Design Concepts and Optimization in Engineering Sciences (ICIDES-II), Penn State Univ., October 26-28, 1987.

[132] **Dulikravich, G. S. (editor)**
Proceedings of the 3rd International Conference on Inverse Design Concepts and Optimization in Engineering Sciences (ICIDES-III), Washington, D.C., October 23-25, 1991.

[133] **Slooff, J. W.**
Computational Methods for Subsonic and Transonic Aerodynamic Design, in: Proc. of 1st Int. Conf. on Inverse Design Concepts and Optimiz. in Eng. Sci. (ICIDES-I), ed. G.S. Dulikravich, Dept. of Aero. Eng. and Eng. Mech., Univ. of Texas, Austin, TX, Oct. 17-18, 1984, pp. 1-68.

[134] **Slooff, J. W. (editor)**
Proceedings of the AGARD Specialist's Meeting on Computational Methods for Aerodynamic Design (Inverse) and Optimization, AGARD CP-463, Loen, Norway, May 2-23, 1989.

[135] **van dem Braembussche, R. (editor)**
Proceedings of a Special Course on Inverse Methods for Airfoil Design for Aeronautical and Turbomachinery Applications, AGARD Report No. 780, Rhode-St.-Genese, Belgium, May 1990.

[136] **Dulikravich, G. S.**

Aerodynamic Shape Design and Optimization: Status and Trends, AIAA Journal of Aircraft, Vol. 29, No. 5, Nov./Dec. 1992, pp. 1020-1026.

[137] **Labrujere, T. E., Slooff, J. W.**
Computational Methods for the Aerodynamic Design of Aircraft Components, in: Annual Review of Fluid Mechanics,Vol. 25, 1993, pp. 183-214.

[138] **Dulikravich, G. S.**
Shape Inverse Design and Optimization for Three-Dimensional Aerodynamics, AIAA invited paper 95-0695, AIAA Aerospace Sciences Meeting, Reno, NV, January 9-12, 1995; also to appear in AIAA Journal of Aircraft.

[139] **Volpe, G.**
Inverse Design of Airfoil Contours: Constraints, Numerical Methods and Applications, AGARD CP-463, Loen, Norway, May 22-23, 1989, Ch. 4.

[140] **Zhu, Z., Sobieczky, H.**
An Engineering Approach for Nearly Shock-Free Wing Design, Proc. of the Internat. Conf. on Fluid Mech., Beijing, China, July 1987.

[141] **Dulikravich, G. S.**
A Criteria for Surface Pressure Specification in Aerodynamic Shape Design, AIAA paper 90-0124, Reno, NV, January 8-11, 1990.

[142] **Huang, C.-Y., Dulikravich, G. S.**
Stream Function and Stream-Function-Coordinate (SFC) Formulation for Inviscid Flow Field Calculations, Computer Methods in Applied Mechanics and Engineering, Vol. 59, November 1986, pp. 155-177.

[143] **Sherif, A., Hafez, M.**
Computation of Three Dimensional Transonic Flows Using Two Stream Functions, Internat. Journal for Numerical Methods in Fluids, Vol. 8, 1988, pp. 17-29.

[144] **Stanitz, J. D.**
A Review of Certain Inverse Methods for the Design of Ducts With 2- or 3-Dimensional Potential Flow, Appl. Mech. Rev., Vol. 41, No. 6, June 1988, pp. 217-238.

[145] **Zhang, S.**
Streamwise Transonic Computations, Ph. D. dissertation, Department of Mathematics & Statistics, University of Windsor, Windsor, Ontario, Canada, December 1993.

[146] **Chen, N.-X., Zhang, F.-X., Dong, M.**
Stream-Function-Coordinate (SFC) Method for Solving 2D and 3D Aerodynamics Inverse Problems of Turbomachinery Flows, Inverse Problems in Engineering, Vol. 1, No. 3, March 1995.

[147] **Garabedian, P., McFadden, G.**
Design of Supercritical Swept Wings, AIAA J., Vol. 20, No. 3, March 1982, pp. 289-291.

[148] **Malone, J., Vadyak, J., Sankar, L. N.**

A Technique for the Inverse Aerodynamic Design of Nacelles and Wing Configurations, AIAA Journal of Aircraft, Vol. 24, No. 1, January 1987, pp. 8-9.

[149] **Hazarika, N.**
An Efficient Inverse Method for the Design of Blended Wing-Body Configurations, Ph.D. Thesis, Aerospace Eng. Dept., Georgia Institute of Technology, June 1988.

[150] **Malone, J. B., Narramore, J. C., Sankar, L. N.**
An Efficient Airfoil Design Method Using the Navier-Stokes Equations, AGARD Specialists' Meeting on Computational Methods for Aerodynamic Design (Inverse) and Optimization, AGARD-CP-463, Loen, Norway, May 22-23, 1989.

[151] **Bristow, D. R., Hawk, J. D.**
Subsonic 3-D Surface Panel Method for Rapid Analysis of Multiple Geometry Perturbations, AIAA paper 82-0993, St. Louis, MO, June 7-11, 1982.

[152] **Fornasier, L.**
An Iterative Procedure for the Design of Pressure-Specified Three-Dimensional Configurations at Subsonic and Supersonic Speeds by Means of a Higher-Order Panel Method, AGARD-CP-463, Loen, Norway, May 22-23, 1989, Ch. 6.

[153] **Kubrynski, K.**
Design of 3-Dimensional Complex Airplane Configurations with Specified Pressure Distribution via Optimization, Proc. of 3rd Int. Conf. on Inverse Design Concepts and Optimiz. in Eng. Sci. (ICIDES-III), editor: Dulikravich, G. S., Washington, D. C., October 23-25, 1991.

[154] **Wang, Z., Dulikravich, G. S.**
Inverse Shape Design of Turbomachinery Airfoils Using Navier-Stokes Equations, AIAA paper 95-0304, Reno. NV, January 9-12, 1995.

[155] **Zannetti, L., di Torino, P., Ayele, T. T.**
Time Dependent Computation of the Euler Equations for Designing Fully 3D Turbomachinery Blade Rows, Including the Case of Transonic Shock Free Design, AIAA paper 87-0007, Reno, NV, January 12-15, 1987.

[156] **Leonard, O.**
Subsonic and Transonic Cascade Design, AGARD-CP-463, Loen, Norway, May 22-23, 1989, Ch. 7.

[157] **Leonard, O., van den Braembussche, R. A.**
Design Method for Subsonic and Transonic Cascade With Prescribed Mach Number Distribution, ASME J. of Turbomachinery, Vol. 114, July 1992, pp. 553-560.

[158] **Takanashi, S.**
Iterative Three-Dimensional Transonic Wing Design Using Integral Equations, J. of Aircraft, Vol. 22, No. 8, August 1985, pp. 655-660.

[159] **Hua, J., Zhang, Z.-Y.**
Transonic Wing Design for Transport Aircraft, ICAS 90-3.7.4, Stockholm, Sweden, September 1990, pp. 1316-1322.

[160] **Bartelheimer, W.**
An Improved Integral Equation Method for the Design of Transonic Airfoils and Wings, AIAA-95-1688-CP, San Diego, CA, June 1995.

[161] **Kaiden, T., Ogino, J., Takanashi, S.**
Non-Planar Wing Design by Navier-Stokes Inverse Computation, AIAA paper 92-0285, Reno, NV, January 6-9, 1992.

[162] **Fujii, K., Takanashi, S.**
Aerodynamic Design Methods for Aircraft and their Notable Applications - Survey of the Activity in Japan, Proc. of 3rd Int. Conf. on Inverse Design Concepts and Optimiz. in Eng. Sci. (ICIDES-III), editor: Dulikravich, G. S., Washington, D.C., October 23-25, 1991.

CHAPTER 11

AERODYNAMIC SHAPE OPTIMIZATION METHODS

G.S. Dulikravich

The Pennsylvania State University, University Park, PA, USA

11.1 Introduction

Although fast and accurate in creating aerodynamic shapes compatible with the specified surface pressure distribution, the inverse shape design methods create configurations that are not optimal even at the design operating conditions [163], [164]. At off design conditions, these configurations often perform quite poorly except when the specified surface pressure distribution, if available at all, would be provided by an extremely accomplished aerodynamicist. When using inverse shape design methods, it is physically unrealistic to generate a 3-D aerodynamic configuration that simultaneously satisfies the specified surface distribution of flow variables, manufacturing constraints (smooth variation of a lifting surface sweep and twist angles, smooth variation of its taper, etc.) and achieves the best global aerodynamic performance (overall total pressure loss minimized, lift/drag maximized, etc.). The designer should use an adequate global optimization algorithm that can utilize any available flow-field analysis code without changes and efficiently optimize the overall aerodynamic characteristics of the 3-D flight vehicle subject to the finite set of desired constraints. The constraints could be purely geometrical or they can be of the overall aerodynamic nature (minimize overall drag for the given values of flight speed, angle of attack and overall lift force, etc.). These objectives can only be met by performing an aerodynamic shape constrained optimization instead of an inverse shape design.

The size and shape of the mathematical space that contains all the design variables (for example, coordinates of all surface points) is very large and complex in a typical 3-D case. To find a global minimum of such a space requires a sophisticated numerical optimization algorithm that avoids local minima, honors the specified constraints and stays within the feasible design domain. The design variable space in a typical aerodynamic shape optimization has a number of local minima. These minima are very hard to escape from even by switching the objective function formulation [165] or consecutive spline fitting and interpolation of the unidirectional search step parameter [166].

There are several fundamental concepts in creating an optimization algorithm. One family of optimization algorithms is based on reducing the objective or cost function (for example, aerodynamic drag) by evaluating the gradient of the cost function and then updating the design variables in the negative gradient direction [167]. Evolution search or genetic algorithms is another family of optimization algorithms that is based on a semi-random sampling through the design variable space and does not require any gradient evaluations [168], [169]. Since both families of optimization algorithms require flow-field analysis to be performed on every perturbed aerodynamic configuration, the optimization of 3-D aerodynamic shapes is a very computationally intensive task.

In a gradient-search optimization approach the flow analysis code must be called at least once for each design variable during each optimization cycle in order to compute the gradient of the objective function if one-sided finite differencing is used for the gradient evaluation. If a more appropriate central differencing is used for the gradient evaluation, the number of calls to the 3-D flow-field analysis code will immediately double. Despite this, the optimization algorithms are still often misused to minimize the difference between the specified and the computed surface flow data in inverse shape design - a task that is significantly more economical when accomplished with any of the standard inverse shape design algorithms.

The most serious drawback of the brute force application of the gradient search optimization in 3-D aerodynamics is that the computing costs increase nonlinearly with the growing number of design variables thus making these algorithms suitable for smaller optimization problems. On the other hand, the computing cost of using evolution search algorithms increases only moderately with the number of design variables (Figure 74) thus making these algorithms more suitable for large optimization problems. Only these optimization algorithms that require minimum number of calls to the flow-field analysis code will be realistic candidates for the 3-D aerodynamic shape optimization.

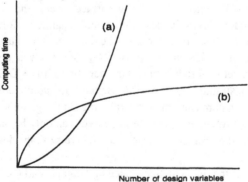

Figure 74   **A sketch of the dependency of computational effort as a function of the number of design variables for the case of a classical gradient search (a) and genetic (b) optimization algorithms.**

Since the actual 3-D flow-field analysis codes of Euler of Navier-Stokes type are very time consuming, the designer is forced to restrict the design space by working with a relatively small number of the design variables for parameterization (fitting polynomials) of either the 3-D surface geometry [170]-[172] or the 3-D surface pressure field [173], [174]. The optimization code then needs to identify the coefficients in these polynomials. The most plausible choices are cubic splines. Chebyshev and Fourier polynomials [175]-[177] are not advisable because they become excessively oscillatory with the increasing number of terms in the polynomial. Moreover, when perturbing any of the coefficients in such a polynomial, the entire 3-D shape will change. Since it is absolutely necessary to constrain and sometimes disallow motion of particular parts of the 3-D surface, the most promising choices for the 3-D parameterization appear to be different types of b-splines [178], [179], [171], [180], local analytical surface patches [181], and local polynomial basis functions [182]. Only when it is possible to use simple and very fast flow-field analysis codes could we afford an ideal optimization situation where each surface grid point on the 3-D optimized configuration is allowed to move independently.

Single-cycle optimization [183] offers one viable approach at reducing the computing costs. Here, the flow-field analysis code is run on each perturbed aerodynamic configuration for only a small number of iterations (instead to a full convergence) before an optimizer is used to determine the new geometry. An optimal aerodynamic shape is then found by optimally weighing each of the number of feasible configurations that can be obtained using inverse design methods. Hence, this optimization approach guarantees that the final configuration will be realistically shaped and manufacturable, although the range of geometric parameters to be optimized is limited by the geometry of the extreme members of the original family of configurations.

11.2 Optimization Using Sensitivity Derivatives

It is often desirable to have a capability to predict the behavior of the inputs to an arbitrary system by relating the outputs to the inputs via a sensitivity derivative matrix [184]-[188], while treating the system as a black box. The sensitivity derivative matrix can be used for the purpose of controlling the system outputs or to achieve an optimized constrained design that depends on the system outputs. The objective is to generate approximations of the infinite dimensional sensitivities and to transfer these approximate derivatives to the optimizer together with the approximate function evaluations. The control variables are then updated with the sensitivity derivatives which are the gradients of the cost function with respect to the control variables. The general concepts for the sensitivity analysis can be summarized as follows [185].

The system of governing flow-field governing equations after discretization results in a system of non-linear algebraic equations

$$R(Q(D), X(D), D) = 0 \qquad (80)$$

where Q is the vector of the solution variables in the flow-field governing system, X is the computational grid and D is the vector of design variables (for example, coordinates of 3-D aerodynamic shape surface points). Hence

$$\frac{dR}{dD} = \frac{\partial R}{\partial Q}\frac{dQ}{dD} + \frac{\partial R}{\partial X}\frac{dX}{dD} + \frac{\partial R}{\partial D} = 0 \tag{81}$$

Similarly, aerodynamic output functions (lift, drag, lift/drag, moment, etc.) are defined as

$$F = F(Q(D), X(D), D) \tag{82}$$

Hence

$$\frac{dF}{dD} = \frac{\partial F}{\partial Q}\frac{dQ}{dD} + \frac{\partial F}{\partial X}\frac{dX}{dD} + \frac{\partial F}{\partial D} \tag{83}$$

System (81) is solved for the sensitivity derivatives of the field variables, dQ/dD, which are then substituted into the system (83) in order to obtain the sensitivity derivatives of the desired aerodynamic outputs, dF/dD. This approach is typically used if the dimension of **F** is greater than that of **D** which is seldom the case in a 3-D aerodynamic shape design.

When the number of design variables **D** is larger than the number of the aerodynamic output functions **F**, it is more economical to avoid solving for dQ/dD. This can be accomplished by using an adjoint operator approach where a linear system

$$\left(\frac{\partial R}{\partial Q}\right)^T A + \left(\frac{\partial F}{\partial Q}\right)^T = 0 \tag{84}$$

must be solved first. Here, **A** is a discrete adjoint variable matrix associated with the aerodynamic output functions, **F**. Substituting equation (84) into equation (83), it follows from equation (81) that the aerodynamic output derivatives of interest can be computed from

$$\frac{dF}{dD} = A^T \cdot \left(\frac{\partial R}{\partial X}\frac{dX}{dD} + \frac{\partial R}{\partial D}\right) + \frac{\partial F}{\partial X}\frac{dX}{dD} + \frac{\partial F}{\partial D} \tag{85}$$

This quasi-analytical approach to computing sensitivity derivatives is more economical and accurate than when evaluating the derivatives using finite differencing. Nevertheless, sensitivity analysis is a very costly process requiring a large number of analysis runs.

In the gradient-search optimization approach the flow analysis code must be called at least once for each design variable in order to compute the gradient of the objective function during each optimization cycle. Since each call to the analysis code is very expensive, such an approach to design is justified only if a small number of design variables is used. In the case of a 3-D design, this is hardly justifiable even if one uses 3-D surface geometry parametrization which severely constrains 3-D optimal configurations.

One of the most promising recent developments in the aerodynamic shape design optimization is a method that treats the entire system of partial differential equations governing the flow-field as constraints, while treating coordinates of all surface grid points as design variables [189]. This approach eliminates the need for geometry parameterization using shape functions to define changes in the geometry. Since fluid dynamic variables, $\mathbf{Q}$, are treated here as the design variables, this method allows for rapid computation of partial derivatives of the objective function with respect to the design variables. This approach is straightforward to comprehend and efficient to implement in Newton-type direct flow analysis algorithms where solutions of the equations for dQ/dD or $\mathbf{A}$ amount to a simple back-substitution. The problem is that the classical Newton iteration algorithm is practically impossible to implement for 3-D aerodynamic analysis codes because of its excessive memory requirements when performing direct LU factorization of the coefficient matrix.

Instead of using an exact Newton algorithm in the flow-analysis code, it is more cost effective to use a quasi-Newton iterative formulation or an incremental iterative strategy [185], [190] given in the form

$$\frac{\partial \Re}{\partial Q} \cdot D\left(\frac{dQ}{dD}\right) = \left(\frac{\partial R}{\partial D}\right)^n = \frac{\partial R}{\partial D} \cdot \left(\frac{dQ}{dD}\right)^n + \frac{\partial R\,dX}{\partial X\,dD} + \frac{\partial R}{\partial D} \tag{86}$$

$$\left(\frac{dQ}{dD}\right)^{n+1} = \left(\frac{dQ}{dD}\right)^n + D\left(\frac{dQ}{dD}\right) \qquad ; n=1,2,3,\dots \tag{87}$$

where $\frac{\partial \Re}{\partial Q}$ could be any fully-converged numerical approximation of the exact Jacobian matrix.

11.3 Constrained Genetic Evolution Optimization

Besides a wide variety of the gradient-based optimization algorithms, truly remarkable results were obtained using an evolution type genetic algorithm or GA [168],[169],[191]. The GA methods simulate the mechanics of natural genetics for artificial systems based on operations which are the counter parts of the natural ones. They rely on the use of a random selection process which is guided by probabilistic decisions. In general, a GA is broken into three major steps: reproduction, crossover, and mutation. An initial population of complete design variable sets are analyzed according to some cost function. Then, this population is merged using a crossover methodology to create a new population. This process continues until a global minimum is found. Generally, the design variable set that corresponds to the minimum point of the cost function will be representative as having the most "successful" features of previous "generations" of designs in the optimization process.

Although the standard genetic algorithm (GA) is computationally quite expensive since it requires a large number of calls to the flow-field analysis code, the robustness of this algorithm and the ease of its implementation have created a recently renewed interest in applying it for aerodynamic shape design [192]-[201]. Nevertheless, the examples presented in these publications involve aerodynamic shape optimization with a small number of design variables that form a relatively compact function space. Solutions of such optimization problems would be considerably more efficient when using more common gradient search algorithms. Moreover, none of the examples in these publications attempt to treat equality-type constrained optimization which represents the most difficult problem for a typical GA algorithm. The classical GA can handle constraints on the design variables, but it is not inherently capable of handling constraint functions [198]. Most of the recent publications involving the GA and aerodynamic shape optimization have involved problems posed in such a way as to eliminate constraint functions, or to penalize the cost function when a constraint is violated. These treatments of constraints reduce the chance of arriving at the global minimum.

Probably the most attractive feature of the GA is its remarkable robustness since it is not a gradient-based search method. The GA is exceptional at avoiding local minima, because it tests possible designs over a large design variable space. Hence, the GA is especially suitable for handling a large number of design variables that belong to widely different engineering disciplines, thus making it particularly suitable for true omplex multidisciplinary coptimization problems. The GA is especially suitable for the types of problems where the sensitivity derivatives might be discontinuous [173] which is sometimes the problem in 3-D aerodynamic optimization. The number of cost function evaluations per design iteration of a GA does not depend on the number of design variables. Rather, it depends on the size of the initial population.

There are some subtleties associated with the GA that, if treated properly, greatly increases the effectiveness and usefulness of the method. One such subtlety involves the crossover procedure. When two members of the population are chosen for a crossover, their design variable sets are generally encoded into strings called "chromosomes". After a crossover has taken place, the "child" variable set is recovered by decoding its newly created chromosome string. When the design variables are floating-point numbers, as is usually the case, this coding and decoding process can introduce a loss of precision arising from numerical truncation on a finite precision computer. This is particularly true when the chromosome string format is defined by a "bit-string" (a base 2 number), and the operating language is FORTRAN. A crossover method that preserves full precision in the design variables can be developed [202] by changing the format of the coded chromosome string and by the implementation of C and C++ as the operating language instead of FORTRAN. These languages allow bitwise shifting of floating point numerics and require no coding or decoding processes whatsoever. The C++ language further allows object-oriented programming techniques that provide a platform for the truest genetic interaction of design variable sets. Equality and non-equality constraints can be incorporated by using Rosen's projection method [202],[180]. This improved GA could be used as a black box optimizer in aerodynamics, elasticity, heat transfer, etc, or in a multidisciplinary design optimization.

11.4 Hybrid GA/Gradient Search Constrained Optimization

The GA has proven itself to be an effective and robust optimization tool for large variable-set problems if the cost function evaluations are very cheap to perform. Nevertheless, in the field of 3-D aerodynamic shape optimization we are faced with the more difficult situation where each cost function evaluation is extremely costly and the number of the design variables is relatively large. Standard GA will require large memory if large number of design variables are used. The number of cost function evaluations per design iteration of a GA increases only mildly with the number of design variables, while increasing rapidly with the increased size of the initial population. Also, the classical GA can handle constraints on the design variables, but it is not inherently capable of handling constraint functions [198]. Thus, the brute force application of the standard GA to 3-D aerodynamic shape design optimization is economically unjustifiable.

Consequently, a hybrid optimization made of a combination of the GA and a gradient search optimization or the GA and an inverse design method has been shown be an advisable way to proceed [180], [201]-[203]. Preliminary results obtained with different versions of a hybrid optimizer that uses a GA for the overall logic, a quasi-Newtonian gradient-search algorithm or a feasible directions method [167] to ensure monotonic cost function reduction, and a Nelder-Mead sequential simplex algorithm or a steepest descent methodology of the design variables into feasible regions from infeasible ones has proven to be effective at avoiding local minima. Since the classical GA does not ensure monotonic decrease in the cost function, the hybrid optimizer could store information gathered by the genetic searching and use it to determine the sensitivity derivatives of the cost function and all constraint functions [203], [180]. When enough information has been gathered and the sensitivity derivatives are known, the optimizer switches to the feasible directions method (with quadratic subproblem) for quickly proceeding to further improve on the best design.

One possible scenario for a hybrid genetic algorithm can be summarized as follows:

- Let the set of population members define a simplex like that used in the Nelder-Mead method.

- If the fitness evaluations for all of the population members does not yield a better solution, then define a search direction as described by the Nelder-Mead method.

- If there are active inequality constraints, compute their gradients and determine a new search direction by solving the quadratic subproblem.

• If there are active equality constraints, project this search direction onto the subspace tangent to the constraints.

• Perform line search.

Our recent work [203], [180] indicates that the hybrid GA can yield answers (Figure 75 and Figure 76) not obtainable by standard gradient methods at comparable convergence rates (Figure 77). This hybrid optimizer can also handle non-linear constraint functions, although the main computational cost will be incurred by enforcing the constraints since this task will involve evaluating the gradients of the constraint functions.

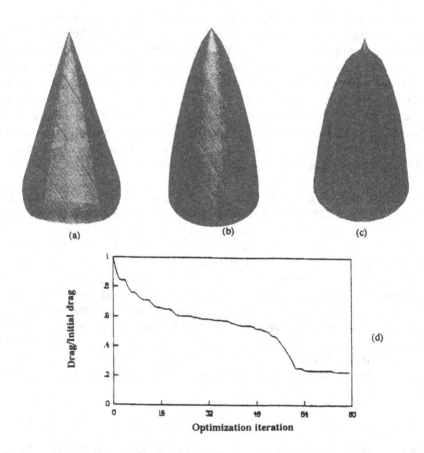

Figure 75   **Our hybrid gradient search/genetic constrained optimizer keeps on improving the design even beyond the known analytic optimal solution (von Karman and Sears-Haack smooth ogives). It transformed an initially conical configuration (a) with 100% drag into a smooth ogive shape (b) with 56% drag and finally into a star-shaped spiked missile (c) with only 22% of the initial drag. Convergence history (d) shows monotonic convergence although 480 variables were optimized with a constrained volume and length of the missile [203], [180].**

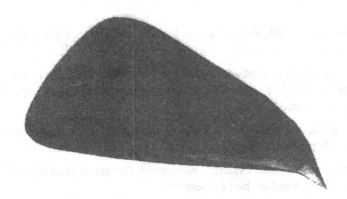

Figure 76   Starting from an initially conical body at zero angle of attack and a hypersonic oncoming flow, the hybrid GA/gradient search constrained algorithm arrived at this smooth, cambered 3-D lifting body with a maximized value of lift-to-drag while maintaining the initial volume and length [203], [180].

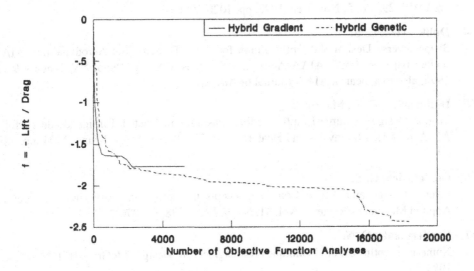

Figure 77   Comparison of convergence histories of a constrained gradient search and a constrained hybrid genetic/gradient search optimizer for example shown in Figure 76, [203], [180].

11.5 Conclusions

From the brief survey of positive and negative features of different optimization algorithms it can be concluded that:

- brute force application of gradient search sensitivity-based optimization methods is very computationally intensive and unreliable for large aerodynamic problems since they are prone to local minima,

- brute force application of genetic evolution optimizers is very computationally intensive for realistic 3-D problems that involve a number of constraints, and

- use of hybrid genetic/gradient search constrained optimization algorithms offers the most reliable performance and the best convergence.

11.6 References

[163] **Dulikravich, G. S.**
Aerodynamic Shape Design and Optimization: Status and Trends, AIAA Journal of Aircraft, Vol. 29, No. 5, Nov./Dec. 1992, pp. 1020-1026.

[164] **Dulikravich, G. S.**
Shape Inverse Design and Optimization for Three-Dimensional Aerodynamics, AIAA invited paper 95-0695, AIAA Aerospace Sciences Meeting, Reno, NV, January 9-12, 1995; also to appear in AIAA Journal of Aircraft.

[165] **Dulikravich, G. S., Martin, T. J.**
Inverse Design of Super-Elliptic Cooling Passages in Coated Turbine Blade Airfoils, AIAA J. of Thermophysics and Heat Transfer, Vol. 8, No. 2, April-June, 1994, pp. 288-294.

[166] **Dulikravich, G. S.**
Inverse Design and Active Control Concepts in Strong Unsteady Heat Conduction, Applied Mechanics Reviews, Vol. 41, No. 6, June 1988, pp. 270-277.

[167] **Vanderplaats, G. N.**
Numerical Optimization Techniques for Engineering Design, McGraw-Hill, New York, 1984.

[168] **Goldberg, D. E.**
Genetic Algorithms in Search, Optimization and Machine Learning, Addison-Wesley, 1989.

[169] **Davis, L.**
Handbook of Genetic Algorithms, Reinhold, 1990.

[170] **Madabhushi, R.K., Levy, R., Pinkus, S. M.**
Design of Optimum Ducts Using an Efficient 3-D Viscous Computational Flow Analysis, Proc. of 2nd Int. Conf. on Inverse Design Concepts and Optimiz. in Eng. Sci. (ICIDES-II), ed. G.S. Dulikravich, Dept. of Aero. Eng., Penn State Univ., University Park, PA, Oct. 26-28, 1987, pp.147-166.

[171] **Huddleston, D. H., Mastin, C. W.**
Optimization of Aerodynamic Designs Using Computational Fluid Dynamics, Proc. of 7th Int. Conf. on Finite Element Meth. in Flow Problems, ed. Chung, T.J. and Karr, G.R., Univ. of Alabama in Huntsville Press, April 3-7, 1989, pp. 899-907.

[172] **Cosentino, G. B.**
Numerical Optimization Design of Advanced Transonic Wing Configurations, AIAA paper 85-0424, Reno, NV, Jan. 14-17, 1985.

[173] **van Egmond, J. A.**
Numerical Optimization of Target Pressure Distributions for Subsonic and Transonic Airfoil Design, Computational Methods for Aerodynamic Design (Inverse) and Optimization, AGARD-CP-463, Loen, Norway, May 22-23, 1989, Ch. 17.

[174] **van den Dam, R. F., van Egmond, J. A., Slooff, J. W.**
Optimization of Target Pressure Distributions, AGARD Report No. 780, Rhode-St.-Genese, Belgium, May 14-18, 1990.

[175] **Dulikravich, G.S., Sheffer, S.**
Aerodynamic Shape Optimization of Arbitrary Hypersonic Vehicles, Proc. of 3rd Int. Conf. on Inverse Design Concepts and Optimiz. in Eng. Sci. (ICIDES-III), ed. G.S. Dulikravich, Washington, D.C., October 23-25, 1991.

[176] **Sheffer, S. G., Dulikravich, G.S.**
Constrained Optimization of Three-Dimensional Hypersonic Vehicle Configurations, AIAA paper 93-0039, AIAA Aerospace Sciences Meeting, Reno, NV, Jan. 11-14, 1993.

[177] **Dulikravich, G. S., Sheffer, S. G.**
Aerodynamic Shape Optimization of Hypersonic Configurations Including Viscous Effects, AIAA 92-2635, AIAA 10th Applied Aerodynamics Conference, Palo Alto, CA, June 22-24, 1992.

[178] **Mortenson, M. E.**
Geometric Modeling, John Wiley & Sons, 1985, pp. 125-134.

[179] **Farin, G.**
Curves and Surfaces for Computer Aided Geometric Design, Second Edition, Academic Press, Boston, 1990.

[180] **Foster, N. F., Dulikravich, G. S., Bowles, J.**
Three-Dimensional Aerodynamic Shape Optimization Using Genetic Evolution and Gradient Search Algorithms, AIAA paper 96-0555, AIAA Aerospace Sciences Meeting, Reno, NV, January 15-19, 1996.

[181] **Sobieczky, H., Stroeve, J. C.**

Generic Supersonic and Hypersonic Configurations, AIAA paper 91-3301-CP, Proc. 9th AIAA Applied Aerodynamics Conference, Baltimore, MD, Sept. 22-24, 1991.

[182] **Kubrynski, K.**
Design of 3-Dimensional Complex Airplane Configurations with Specified Pressure Distribution via Optimization, Proc. of 3rd Int. Conf. on Inverse Design Concepts and Optimiz. in Eng. Sci. (ICIDES-III), editor: Dulikravich, G. S., Washington, D. C., October 23-25, 1991.

[183] **Rizk, M. H.**
Aerodynamic Optimization by Simultaneously Updating Flow Variables and Design Parameters, AGARD CP-463, Loen, Norway, May 22-23, 1989.

[184] **Tortorelli, D. A., Michaleris, P.**
Design Sensitivity Analysis: Overview and Review, Inverse Problems in Engineering, Vol. 1, No. 1, 1994, pp. 71-105.

[185] **Taylor, A. C., Newman, P. A., Hou, G. J.-W., Jones, H. E.**
Recent Advances in Steady Compressible Aerodynamic Sensitivity Analysis, IMA Workshop on Flow Control, Institute for Math. and Appl., Univ. of Minnesota, Minneapolis, MN, November 1992.

[186] **Elbanna, H. M., Carlson, L. A.**
Determination of Aerodynamic Sensitivity Coefficients Based on the Three-Dimensional Full Potential Equation, AIAA paper 92-2670, June 1992.

[187] **Baysal, O., Eleshaky, M. I.**
Aerodynamic Design Optimization Using Sensitivity Analysis and Computational Fluid Dynamics, AIAA paper 91-0471, Reno, NV, Jan. 7-10, 1991.

[188] **Korivi, V. M., Taylor, A. C., Hou, G. W., Newman, P. A., Jones, H.**
Sensitivity Derivatives for Three-Dimensional Supersonic Euler Code Using Incremental Iterative Strategy, AIAA Journal, Vol. 32, No. 6, June 1994, pp. 1319-1321..

[189] **Felker, F.**
Calculation of Optimum Airfoils Using Direct Solutions of the Navier-Stokes Equations, AIAA paper 93-3323-CP, Orlando, FL, July 1993.

[190] **Huddleston, D. H, Soni, B. K., Zheng, X.**
Application of a Factored Newton-Relaxation Scheme to Calculation of Discrete Aerodynamic Sensitivity Derivatives, AIAA paper 94-1894, Colorado Springs, CO, June 1994.

[191] **Hajela, P.**
Genetic Search - An Approach to the Nonconvex Optimization Problem, AIAA Journal, Vol. 28, No. 7, 1990, pp. 1205-1210.

[192] **Misegades, K. P.**
Optimization of Multi-Element Airfoils, Von Karman Institute for Fluid Dynamics, Belgium, Project Report 1980-5, June 1980.

[193] **Gregg, R. D., Misegades, K. P.**
Transonic Wing Optimization Using Evolution Theory, AIAA paper 87-0520, January 1987.

[194] **Ghielmi, L., Marazzi, R., Baron, A.**
A Tool for Automatic Design of Airfoils in Different Operating Conditions, AGARD-CP-463, Loen, Norway, May 22-23, 1989.

[195] **Rocchetto, A., Poloni, C.**
A Hybrid Numerical Optimization Technique Based on Genetic and Feasible Direction Algorithms for Multipoint Helicopter Rotor Blade Design, Twenty First European Rotorcraft Forum, St. Petersburg, Russia, 1995, Paper No II7,1-19.

[196] **Gage, P., Kroo, I.**
A Role for Genetic Algorithms in a Preliminary Design Environment, AIAA paper 93-3933, August 1993.

[197] **Mosetti, G., Poloni, C.**
Aerodynamic Shape Optimization by Means of a Genetic Algorithm, Proceedings of the 5th International Symposium on Computational Fluid Dynamics, (editor: H. Daiguji), Sendai, Japan, Japan Society for CFD, Vol. II, 1993, pp. 279-284.

[198] **Crispin, Y.**
Aircraft Conceptual Optimization Using Simulated Evolution, AIAA paper 94-0092, Reno, NV, January 10-13, 1994.

[199] **Quagliarella, D., Cioppa, A. D.**
Genetic Algorithms Applied to the Aerodynamic Design of Transonic Airfoils, AIAA paper 94-1896, Colorado Springs, CO, June 1994.

[200] **Yamamoto, K., Inoue, O.**
Applications of Genetic Algorithm to Aerodynamic Shape Optimization, AIAA CFD Conference Proceedings, San Diego, CA, June 1995, AIAA-95-1650-CP, pp. 43-51.

[201] **Dervieux, A., Male, J.-M., Marco, N., Periaux, J., Stoufflet, B., Chen, H. Q., Cefroui, M.**
Numerical vs. Non-Numerical Robust Optimisers for Aerodynamic Design Using Transonic Finite-Element Solvers, AIAA CFD Conference Proceedings, San Diego, CA, June 1995, AIAA-95-1761-CP, pp. 1339-1347.

[202] **Poloni, C., Mosetti, G.**
Aerodynamic Shape Optimization by Means of Hybrid Genetic Algorithm, Proc. 3rd Internat. Congress on Industrial and Appl. Math., Hamburg, Germany, July 3-7, 1995.

[203] **Foster, N. F.**
Shape Optimization Using Genetic Evolution and Gradient Search Constrained Algorithms, M. Sc. thesis, Dept. of Aerospace Eng., The Pennsylvania State University, University Park, PA, August 1995.

CHAPTER 12

COMBINED OPTIMIZATION AND INVERSE DESIGN
OF 3-D AERODYNAMIC SHAPES

G.S. Dulikravich

The Pennsylvania State University, University Park, PA, USA

12.1 Introduction

The main drawback of using constrained optimization in 3-D aerodynamic shape design is that it requires anywhere from hundreds to tens of thousands of calls to a 3-D flow-field analysis code. Since certain general 3-D aerodynamic shape inverse design methodologies require only a few calls to a modified 3-D flow-field analysis code, it would be highly desirable to create a hybrid new design algorithm that would combine some of the best features of both approaches while requiring less computing time than a few dozen calls to the 3-D flow-field analysis code. We will discuss two such hybrid design formulations that have been proven to work and are distinctly different from each other.

12.2 Target Pressure Optimization Followed by an Inverse Design

The unique feature of this concept [204], [205] is that it offers the most economical approach to constrained aerodynamic shape optimization. It consists of two steps: surface pressure constrained optimization followed by an inverse shape design. This approach avoids most of the limitations of the inverse shape design while requiring considerably less computing time than the direct geometry optimization.

Surface pressure optimization phase of this design approach starts by parameterizing the initial surface pressure distribution using β-splines [206]. Typically, between five and ten control points in the β-spline representation are sufficient to describe the pressure distribution on the upper surface and the similarly small number of control points is sufficient for representation of the pressure distribution on the lower surface of a two-dimensional (2-D) section of the 3-D object. For example, if 7 control points are used to parameterize the upper surface pressure distribution and 7 control points are used to parameterize the lower surface pressure distribution (Figure 78), then 4 control points will have to be used to fix the locations of leading and trailing edge stagnation points. Constraints can be introduced at this stage by specifying the slopes of the pressure distribution at the leading edge. This will require fixing one additional control point on the upper and the lower surface pressure distribution [204]. The steeper the leading edge pressure distribution variation, the larger the leading edge radius of the resulting 2-D aerodynamic section will be. The optimization of the constrained surface pressure distribution can then be achieved [207], [208] by optimizing the 8 "floating" control points each defined by its x-coordinate and the corresponding value of pressure. The optimization is thus performed on surface pressure distribution, not on the actual aerodynamic shape.

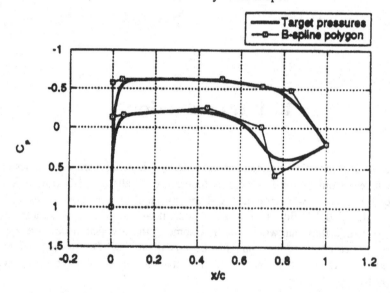

Figure 78   **An example of surface pressure coefficient discretization using β-splines with enforced constraints on the slopes of the Cp curves at the leading edge stagnation point [204].**

This means that the perturbations introduced in the surface pressure distribution during the optimization will have to be somehow related to the global objectives like aerodynamic lift, drag, etc. These global aerodynamic parameters depend on the geometric shape of the object and on the pressure distribution on its surface. It is therefore necessary to relate the geometry of the object and the surface pressure distribution. This is precisely what a typical flow-field analysis code should do except for the fact that geometry of the object is not known yet. A possibility would be to run an inverse shape design code for each surface pressure perturbation and then to integrate the surface pressure in order to evaluate the corresponding lift, drag, etc. This is obviously an unacceptably, inefficient approach.

Instead, the guessed surface pressure distribution can be optimized without ever calling a flow-field analysis code or the inverse shape design code. To accomplish this, use is made of extremely efficient approximate relations [209], [210] relating the integrated surface pressure distribution and the object's maximum relative thickness [209], specified location of the flow transition points, and the aerodynamic drag with any of the classical boundary layer solutions [210]. Then, the optimization of the few coefficients, for example, of β-spline parameterization of the surface pressure distribution curves on each 2-D section can be performed reliably and economically using an optimization algorithm [211], [204], [212]. Optimization of the surface pressure distribution parameters, instead of directly optimizing the 3-D geometry parameters, has several advantages: the total number of optimization variables is reduced, the range of β-spline coefficient variations can be large, the design space does not have an excessive number of local minima, and the sensitivity derivatives do not have discontinuities. This procedure easily accepts constraints on surface pressure distribution such as the desired slopes of the curves at the leading and trailing edges, the maximum value of negative coefficient of pressure, a condition that the surface pressure distribution curves on the upper and the lower surface never cross each other thus avoiding locally negative lift force. These optimization tasks can be reliably achieved using a genetic evolution optimization algorithm [204], although an even more reliable and computationally faster approach would be to use a hybrid genetic/gradient search optimization algorithm [212], [213].

The second phase of this combined 3-D shape design procedure utilizes the optimized surface pressure distribution and any of the fast inverse shape design algorithms [214], [204], [215] to find the corresponding 3-D configuration (Figure 79).

The entire two-step design algorithm requires a minimum development time since β-spline discretization codes, constrained optimization codes, flow-field analysis codes, and inverse design codes are available to most aerodynamicists. If the optimized surface pressure distribution results in a 3-D geometry which violates some of the local geometric constraints, the computed pressure at the constrained points can be used instead of the optimized local pressure. Moreover, this approach gives the designer a partial control of the key elements of the design by asking him to specify a few key bounding features of the "good" surface pressure distribution and then letting the robust and fast optimizer find the details of the optimal pressure distribution. Using this approach to shape optimization the designer also has the ability to visually control and terminate or restart the design process if he decides that the values of some of his specified constraints could be improved.

The method offers the most economical and robust approach to 3-D constrained aerodynamic shape optimization (Figure 80) that consumes approximately 5 - 10 times the computing time needed by a single 3-D flow-field analysis [204]. This is much more cost-effective than the classical approach of simultaneously optimizing the surface pressure distribution and the corresponding 3-D geometry which costs an equivalent of hundreds and even thousands of flow field analysis runs.

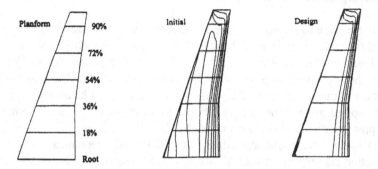

Figure 79   An example of an initial and an optimized surface pressure distribution on a
3-D transonic wing planform. Notice that the optimized pressure distribution
is of the flat "roof-top" type [204].

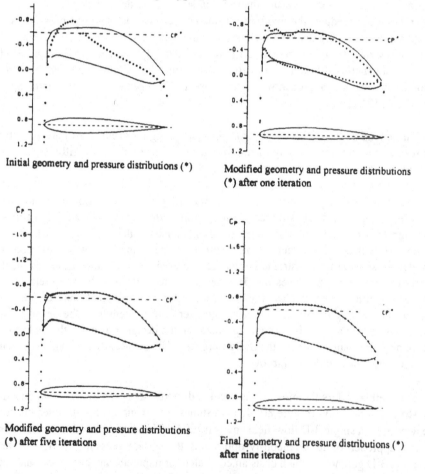

Figure 80   An example of a convergence history of target pressure optimization followed
by transpiration inverse design [204].

12.3 Control Theory (Adjoint Operator) Approach

With the increasing number of variables that need to be optimized, the discrete function space containing the optimization variables tends towards a continuous function space for which it is possible to use a global analytical formulation (an adjoint system) instead of the local discretized formulation (gradient search and genetic evolution algorithms). The control theory (adjoint operator) concept is essentially a gradient search optimization approach where the gradient information is obtained by formulating and solving a set of adjoint partial differential equations rather than evaluating the derivatives using finite differencing. Like the classical optimization algorithms, the control theory (adjoint operator) formulation can be used either for optimizing a 3-D aerodynamic shape by maximizing its lift, minimizing drag, etc., or it can be used as a tool to enforce the desired surface pressure distribution in an inverse shape design process. If the governing system of partial differential equations is large and the solution space is relatively smooth, then the control theory (adjoint operator) approach is more appropriate than the classical optimization approaches [216]-[218]. The mathematical formulation is very involved and was only briefly sketched in the previous lecture.

There have been several approaches at creating an aerodynamics control theory (adjoint operator) formulation [216]-[233]. The early efforts [216]-[220] were highly mathematical, hard to understand and computationally intensive. Since then, two similar approaches have been followed by most researchers. They are associated with the research teams of Professor A. Jameson [221]-[225] who adopted and extended the previous efforts by the French researchers, and the research team of Professor V. Modi [226]-[228] who followed a more comprehensible mechanistic approach practised by researchers in the general fields of inverse shape design in elasticity and heat conduction.

To date, successful and impressive results have been obtained by both research groups where Jameson's group focused on transonic inviscid flow model and Modi's group focused on incompressible viscous laminar and turbulent flow models of the flow field. Published results of 3-D elbow diffuser and airfoil shape optimization using the adjoint operator approach and incompressible laminar and turbulent flow Navier-Stokes equations [226]-[228] suggest that a typical optimized design consumes the amount of computing time that is equivalent to between 20 - 40 flow-field analysis (Figure 81, Figure 82). Similar total effort (an equivalent of 30 - 60 flow-field analysis) was reported for 3-D transonic isolated wing design using Euler equations [21, 22]. This could be compared with several hundreds and even thousands of flow analysis runs when using a typical genetic algorithm or a typical gradient search optimization algorithm. These results dispel earlier reservations [30] that adjoint operator approach formulations might not be computationally efficient since they involve the solution of the governing flow-field equations, an additional set of adjoint equations, and several more interface partial differential equations.

Figure 81   History of the iterative evolution of a minimum drag airfoil starting from a
NACA0018 airfoil at Re = 5000. Labeling numbers correspond to the
iterative cycles each consisting of one flow-field analysis and one solution of
the adjoint system [228].

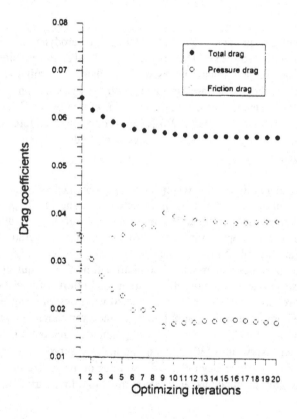

Figure 82   History of the iterative evolution of hydrodynamic drag coefficient during the
shape optimization process with initial shape chosen as NACA 0018 and Re =
5000 [228].

Although very impressive and mathematically involved, the general control theory (adjoint operator) approach has certain problems. One drawback is that it does not always allow for flow separation. The method also suffers from tendency to converge to any of the numerous local minima like most of the gradient search optimizers. An additional drawback is that it requires the derivation of an entirely new system of partial differential equations in terms of some non-physical adjoint variables and specification of their boundary conditions. Choosing correct boundary conditions for the adjoint system is quite a challenge since the adjoint variables are not physical flow quantities [228]. There are many ways to derive the adjoint system and some additional partial differential equations coupling the original system and the adjoint system [221]-[226]. If the adjoint system of partial differential equations is different from the original system of the flow-field governing equations, a significant effort needs to be invested in separately coding the two systems. If the adjoint system is almost the same as the original governing system, the numerical algorithms for the two systems are practically the same and the entire approach could be implemented more readily using the existing CFD analysis software [228]. Actually, if the adjoint system has the same form as the flow-field governing system except that the sign of the convective term in the adjoint system is negative, the solution of the adjoint system will represent a flow in the reverse direction (Figure 83 and Figure 84).

The complexity of the entire adjoint operator formulation makes it difficult to comprehend, implement and modify. Moreover, the adjoint operator formulation is very specific and different formulation needs to be developed and coded for each flow field model (Euler, parabolized Navier-Stokes, Navier-Stokes, etc.). The control theory (adjoint operator) approach is very attractive if the designers want to use only one specific flow-field analysis code as the basis for a design code and if they want to perform design in only one discipline (for example, aerodynamic shape design only). Since the designers use a variety of progressively more sophisticated design codes during the design process and since the design objectives are inherently multidisciplinary, the control theory (adjoint operator) approach can hardly be justified in the context of the multidisciplinary objectives, funds typically available, and the time limits imposed on the designer.

In the case of a truly multidisciplinary problem, a new adjoint system would have to be derived and coded to solve several adjoint systems simultaneously. Since each disciplinary analysis and adjoint system usually has vastly different time scales, the combined multidisciplinary analysis and adjoint system would have a slow convergence rate and an overall marginal stability because of a very large number of local minimums. Reliability of such a design system might be questionable since it is known that sensitivity derivatives for highly non-linear systems might be discontinuous [207]. The adjoint system must be discretized for an approximation to the gradient to be found. This approach is less reliable than the implicit function theorem approach [231], [232]. Therefore, finding the gradients of the objective function using information from the discretized flow-field governing equations is more reliable [232] than if an analytic expression for the gradient is derived in terms of the exact flow-field solution and the solution to the adjoint system [223].

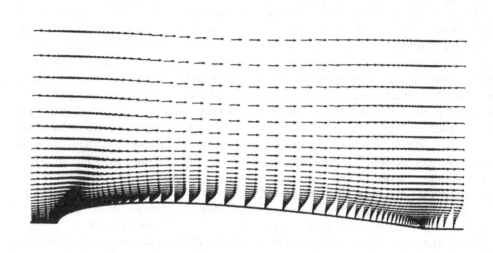

Figure 83   Horizontal components of the fluid velocity vector at different axial locations; the flow is from left to right [228].

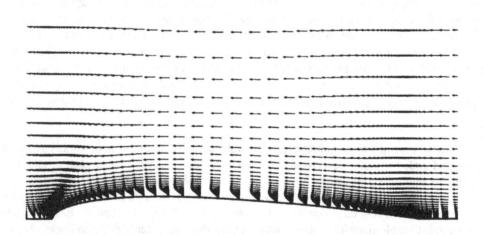

Figure 84   Horizontal components of an adjoint variable at different axial locations; the "flow" of the adjoint vector is in the opposite direction to the physical fluid flow if the adjoint system is made to resemble the flow-field governing system except for the change in sign of the convective term [228].

An attractive extension of the control theory (adjoint operator) approach represents the "one-shot" method [234]-[237]. It implicitly combines a multigrid solution technique with the classical control theory for systems of partial differential equations. Consequently, this approach is faster than the regular control theory (adjoint operator) approach where the multigrid technique was used only to accelerate the flow-field analysis code [225]. Nevertheless, the "one-shot" formulation and implementation are even more complex than the control theory (adjoint operator) formulation, and it suffers from similar reliability issues due to the fact that it might be equally prone to the local minima.

12.4 Conclusions

A number of existing and emerging concepts and methodologies applicable to automatic inverse design and optimization of arbitrary realistic 3-D configurations have recently been surveyed and compared [235]. These attempts to classify the design methods and to expose their major advantages and disadvantages resulted in the following conclusions:

- control theory (adjoint operator) algorithms offer very economical shape design optimization although they are complex to understand and develop, hard to modify, too field-specific, and prone to local minima,

- one-shot method should be further researched as an even more economical possible successor to the adjoint operator formulations with all of its drawbacks, and

- combination of hybrid genetic/gradient search constrained optimization of surface pressure distribution followed by transpiration inverse design offers an attractive approach in the immediate future because of its unsurpassed robustness and acceptable computing cost.

12.5 References

[204] **Obayashi, S., Takanashi, S.**
Genetic Optimization of Target Pressure Distributions for Inverse Design Methods, AIAA Paper 95-1649, San Diego, CA, June 19-22, 1995.

[205] **Goel, S., Lamson, S.**
Automating the Design Process for 3D Turbine Blades, ASME WAM'95, Nov. 14-17, 1995, in Symp. for Design & Optimization (editor: O. Baysal), ASME Vol. 232, pp. 85-95.

[206] **Rogers, D. F., Adams, J. A.**
Mathematical Elements for Computer Graphics, Second Edition, McGraw-Hill, Inc., 1990.

[207] **van Egmond, J. A.**
Numerical Optimization of Target Pressure Distributions for Subsonic and Transonic Airfoil Design, AGARD-CP-463, Loen, Norway, May 22-23, 1989, Ch. 17.

[208] **van den Dam, R. F., van Egmond, J. A., Slooff, J. W.**
Optimization of Target Pressure Distributions, AGARD Report No. 780, Rhode-St.-Gen-

ese, Belgium, May 14-18, 1990.

[209] **Inger, G. R.**
Application of Oswatitsch's Theorem to Supercritical Airfoil Drag Calculation, J. of Aircraft, Vol. 30, No. 3. May-June 1993, pp. 415-416.

[210] **Young, A. D.**
Boundary Layers, AIAA Education Series, 1989.

[211] **Davis, L.**
Handbook of Genetic Algorithms, Reinhold, 1990.

[212] **Foster, N. F., Dulikravich, G. S., Bowles, J.**
Three-Dimensional Aerodynamic Shape Optimization Using Genetic Evolution and Gradient Search Algorithms, AIAA paper 96-0555, AIAA Aerospace Sciences Meeting, Reno, NV, January 15-19, 1996.

[213] **Poloni, C., Mosetti, G.**
Aerodynamic Shape Optimization by Means of Hybrid Genetic Algorithm, Proc. 3rd Internat. Congress on Industrial and Appl. Math., Hamburg, Germany, July 3-7, 1995.

[214] **Takanashi, S.**
Iterative Three-Dimensional Transonic Wing Design Using Integral Equations, J. of Aircraft, Vol. 22, No. 8, August 1985, pp. 655-660.

[215] **Wang, Z., Dulikravich, G. S.**
Inverse Shape Design of Turbomachinery Airfoils Using Navier-Stokes Equations, AIAA paper 95-0304, Reno. NV, January 9-12, 1995.

[216] **Lions, J. L.**
Optimal Control of Systems Governed by Partial Differential Equations, Springer-Verlag, New York, 1971. Translated by S. K. Mitter.

[217] **Angrand, F.**
Optimum Design for Potential Flows, International Journal for Numerical Methods in Fluids, Vol. 3, 1983, pp. 265-282.

[218] **Abergel, F., Temam, R.**
On Some Control Problems in Fluid Mechanics, Theoretical and Comput. Fluid Dynamics, Vol.1, 1990, pp. 303-325.

[219] **Gu, C.-G., Miao, Y.-M.**
Blade Design of Axial-Flow Compressors by the Method of Optimal Control Theory-- Physical Model and Mathematical Expression, ASME paper 86-GT-183 presented at the 31st Internat. Gas Turbine Conf. and Exhibit, Dusseldorf, Germany, June 1986.

[220] **Gu, C.-G., Miao, Y.-M.**
Blade Design of Axial-Flow Compressors by the Method of Optimal Control Theory-- Application of Pontryagin's Maximum Principles, a Sample Calculation and Its Results, ASME paper 86-GT-182 presented at the 31st Internat. Gas Turbine Conf. and Exhibit, Duesseldorf, Germany, June 1986.

[221] **Jameson, A.**
Aerodynamic Design Via Control Theory, J. of Scientific Computing, Vol. 3, 1988, pp. 233-260.

[222] **Jameson, A.**
Optimum Aerodynamic Design via Boundary Control, AGARD-FDP-VKI Special Course, VKI, Rhode St.-Genese, Belgium, April 25-29, 1994.

[223] **Reuther, J., Jameson, A.**
Control Theory Based Airfoil Design for Potential Flow and a Finite Volume Discretization, AIAA paper 94-0499, Reno, NV, January 1994.

[224] **Reuther, J., Jameson, A.**
Control Based Airfoil Design Using the Euler Equations, AIAA paper 94-4272-CP, 1994.

[225] **Jameson, A.**
Optimum Aerodynamic Design Using CFD and Control Theory, AIAA CFD Conference Proceedings, San Diego, CA, June 1995, AIAA-95-1729-CP, pp. 926-949.

[226] **Cabuk, H., Modi, V.**
Optimum Design of Oblique Flow Headers, ASHRAE Winter Meeting, New York, NY, January 19-23, 1991.

[227] **Cabuk, H., Sung, C.-H., Modi, V.**
Adjoint Operator Approach to Shape Design for Internal Incompressible Flows, Proc. of the 3rd Internat. Conf. on Inverse Design Concepts and Opt. in Eng. Sci. (ICIDES-III), Ed.: G.S. Dulikravich, Washington, D.C., Oct. 23-25, 1991, pp. 391-404.

[228] **Huan, J., Modi, V.**
Optimum Design of Minimum Drag Bodies in Incompressible Laminar Flow Using a Control Theory Approach, Inverse Problems in Engineering, Vol. 1, No. 1, 1994, pp. 1-25.

[229] **Zhang, J., Chu, C. K., Modi, V.**
Design of Plane Diffusers in Turbulent Flow, Inverse Problems in Engineering, Vol. 2, No. 2, 1995, pp. 85-102.

[230] **Huan, J., Modi, V.**
Design of Minimum Drag Bodies in Incompressible Laminar Flow, Inverse Problems in Engineering, in press 1996.

[231] **Frank. P. D., Shubin, G. R.**
A Comparison of Optimization-Based Approaches for a Model Computational Aerodynamics Design Problem, Journal of Computational Physics, Vol. 98, 1992, pp. 74-89.

[232] **Dixon, A. E., Fletcher, C. A. J.**
Optimization Applied to Aerofoil Design, Proc. 5th Int. Symp. on CFD, Sendai, Japan, 1993.

[233] **Dulikravich, G. S.**
Aerodynamic Shape Design and Optimization: Status and Trends, AIAA Journal of Aircraft, Vol. 29, No. 5, Nov./Dec. 1992, pp. 1020-1026.

[234] **Dulikravich, G. S.**
Shape Inverse Design and Optimization for Three-Dimensional Aerodynamics, AIAA invited paper 95-0695, AIAA Aerospace Sciences Meeting, Reno, NV, January 9-12, 1995; also to appear in AIAA Journal of Aircraft.

[235] **Ta'asan, S.**
One Shot Methods for Optimal Control of Distributed Parameter Systems I: Finite Dimensional Control, ICASE Report No. 91-2, January 1991.

[236] **Arian, E., Ta'asan, S.**
Multigrid One Shot Methods for Optimal Control Problems: Infinite Dimensional Control, ICASE Report No. 94-52, July 1994.

[237] **Ta'asan, S.**
Trends in Aerodynamics Design and Optimization: A Mathematical Viewpoint, AIAA CFD Conference Proceedings, San Diego, CA, June 1995, AIAA-95-1731-CP, pp. 961-970.

CHAPTER 13

THERMAL INVERSE DESIGN AND OPTIMIZATION

G.S. Dulikravich

The Pennsylvania State University, University Park, PA, USA

13.1 Introduction

During the design of high speed flight vehicles the designer should take into account the aerodynamic heating due to surface friction and the high temperature air behind the strong shock waves. The allowable exterior surface temperatures are limited by the material properties of the skin material of the flight vehicle. In addition, the amount of heat that penetrates the skin structure should be minimized since it will have to be absorbed by the fuel and not allowed to enter the passenger cabin. A typical remedy is to cool the structure by pumping a cooling fluid (typically the fuel) through numerous passages manufactured inside the outer structure of the flight vehicle. A design optimization method should therefore provide the designer with a tool to guide the development of innovative designs of internally cooled, thermally coated or non-coated structures that will cost less to manufacture, have a longer life span, be easier to repair, and sustain higher surface temperatures.

13.2 Determination of Number, Sizes, Locations, and Shapes of Coolant Flow Passages

During the past dozen years, the author's research team has been developing a unique inverse shape design methodology and accompanying software which allows a thermal system designer to determine the minimum number and correct sizes, shapes, and locations of coolant passages in arbitrarily-shaped internally-cooled configurations [238]-[255]. The designer needs to specify both the desired temperatures and heat fluxes on the hot surface, and either temperatures or convective heat coefficients on the guessed coolant passage walls. The designer must also provide an initial guess of the total number, sizes, shapes, and locations of the coolant flow passages. Afterwards, the design process uses a constrained optimization algorithm to minimize the difference between the specified and computed hot surface heat fluxes by automatically relocating, resizing, reshaping and reorienting the initially-guessed coolant passages. All unnecessary coolant flow passages are automatically reduced to a very small size and eliminated while honoring the specified minimum distances between the neighboring passages and between any passage and the thermal barrier coating if such exists.

This type of computer code is highly economical, reliable and geometrically flexible if it utilizes the boundary element method (BEM) instead of finite element or finite difference method for the thermal field analysis. The BEM does not require generation of the interior grid and it is non-iterative [238], [239]. Thus the method is computationally efficient and robust. The resulting shapes of coolant passages are smooth, and easily manufacturable.

The methodology has been successfully demonstrated on 2-D coated and non-coated turbine blade airfoils [240]-[249], scramjet combustor struts [252], and 3-D coolant passages in the walls of 3-D rocket engine combustion chambers [253] and 3-D turbine blades [254], [255].

Nonlinear BEM algorithms are the best choice for the thermal analysis because of their computational speed, reliability (due to their non-iterative nature) and accuracy with elliptic type problems. A simple method for escaping local minima has been implemented and involves switching the objective function when a stationary point is achieved [245]. An accurate method, based on exponential spline fitting and interpolation of the cost function values, has been developed for finding the value of optimal search step parameter during gradient search optimization [244]. It is also possible to develop a version of the 3-D inverse shape design code that will allow for multiple realistically shaped coolant flow passages with an arbitrary number of fins or ribs in each of the passages [252] and prespecified locations of struts [242]. In addition, this version of the 3-D inverse design code could allow for a variable thickness, segmented and non-segmented thermal barrier coatings with temperature-dependent thermal conductivities.

13.3 Determination of Steady Thermal Boundary Conditions

Inverse determination of unknown steady thermal boundary conditions when temperature and heat flux data are not available on certain boundaries is an ill-posed problem [256]-[264] (Figure 85). In this case, additional overspecified measurement data involving both temperature and heat flux are required on some other, more accessible boundaries or at a finite number of points within the domain. For example, when using a BEM algorithm, if at all four vertices designated with subscripts 1, 2, 3, and 4 of a quadrilateral computational grid cell the heat sources pi are known, at two vertices both temperature $\Theta = \overline{\Theta}$ and heat flux $q = \bar{q}$ are known, while at the remaining two vertices neither **Q** or q is known, the boundary integral equation becomes

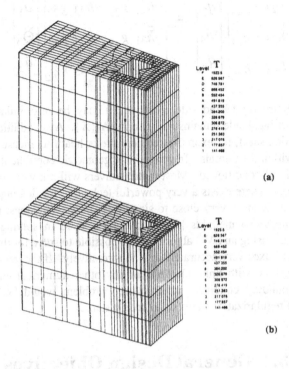

(a)

(b)

Figure 85 Surface isotherms predicted on a circumferentially-periodic three-dimensional segment of a rocket chamber wall with a cooling channel made of four different materials: a) direct problem, b) inverse boundary condition problem [261].

$$
\begin{bmatrix}
\tilde{h}_{11} & \tilde{h}_{12} & \tilde{h}_{13} & \tilde{h}_{14} \\
\tilde{h}_{21} & \tilde{h}_{22} & \tilde{h}_{23} & \tilde{h}_{24} \\
\tilde{h}_{31} & \tilde{h}_{32} & \tilde{h}_{33} & \tilde{h}_{34} \\
\tilde{h}_{41} & \tilde{h}_{42} & \tilde{h}_{43} & \tilde{h}_{44}
\end{bmatrix}
\begin{bmatrix}
\overline{\Theta}_1 \\
\Theta_2 \\
\overline{\Theta}_3 \\
\Theta_4
\end{bmatrix}
=
\begin{bmatrix}
g_{11} & g_{12} & g_{13} & g_{14} \\
g_{21} & g_{22} & g_{23} & g_{24} \\
g_{31} & g_{32} & g_{33} & g_{34} \\
g_{41} & g_{42} & g_{43} & g_{44}
\end{bmatrix}
\begin{bmatrix}
\overline{q}_1 \\
q_2 \\
\overline{q}_3 \\
q_4
\end{bmatrix}
+
\begin{bmatrix}
p_1 \\
p_2 \\
p_3 \\
p_4
\end{bmatrix}
\qquad (88)
$$

Here, coefficients of [h] and [g] matrices are all known because they depend on geomet ric relations and the configuration is known. In order to solve this set, all of the unknowns will be collected on the right-hand side, while all of the knowns are assembled on the left. A simple algebraic manipulation yields the following set:

$$
\begin{bmatrix}
\tilde{h}_{11} & -g_{12} & \tilde{h}_{13} & -g_{14} \\
\tilde{h}_{21} & -g_{22} & \tilde{h}_{23} & -g_{24} \\
\tilde{h}_{31} & -g_{32} & \tilde{h}_{33} & -g_{34} \\
\tilde{h}_{41} & -g_{42} & \tilde{h}_{43} & -g_{44}
\end{bmatrix}
\begin{bmatrix}
\overline{\Theta}_2 \\
q_2 \\
\overline{\Theta}_4 \\
q_4
\end{bmatrix}
=
\begin{bmatrix}
-\tilde{h}_{11} & g_{11} & -\tilde{h}_{13} & g_{13} \\
-\tilde{h}_{21} & g_{21} & -\tilde{h}_{23} & g_{23} \\
-\tilde{h}_{31} & g_{31} & -\tilde{h}_{33} & g_{33} \\
-\tilde{h}_{41} & g_{41} & -\tilde{h}_{43} & g_{43}
\end{bmatrix}
\begin{bmatrix}
\overline{\Theta}_1 \\
\bar{q}_1 \\
\overline{\Theta}_3 \\
\bar{q}_3
\end{bmatrix}
+
\begin{bmatrix}
P_1 \\
P_2 \\
P_3 \\
P_4
\end{bmatrix}
\tag{89}
$$

Since the entire right-hand side is known, it may be reformulated as a vector of knowns, $\{F\}$. The left-hand side remains in the form $[A]\{X\}$. Also, additional equations may be added to the equation set if, for example, temperature or heat flux measurements are known at certain locations within the domain. In general, the geometric coefficient matrix $[A]$ will be non-square and highly ill-conditioned. Most matrix solvers will not work well enough to produce a correct solution. There exists a very powerful technique for dealing with sets of equations that are either singular or very close to singular. These techniques, known as Singular Value Decomposition (SVD) methods [265], are widely used in solving most linear least squares problems. Thus, using an SVD algorithm it is possible to solve for the unknown surface temperatures and heat fluxes very accurately and non-iteratively. If the general formulation of the problem is to be done with finite elements or any other numerical technique instead of BEM, the inverse boundary condition determination problem would become iterative and would require special regularization methods to keep it stable.

13.4 General Design Objectives

The objective of a conjugate heat transfer design and optimization program should be to provide industry with a modular, reliable and proven design optimization tool that will take into account the interaction of 3-D exterior hot gas aerodynamics, heat convection at the hot exterior surface, heat conduction throughout the structure material, and heat convection on the walls of the interior coolant passages. Most of these concepts have already been individually developed, proven and published by the author and other researchers, thus providing strong assurance that the overall integration can be successfully accomplished.

The entire program should be conveniently broken into individual self-sustaining modular tasks. To remain within the desired time frame and budget, the designers should try to utilize as much as possible the existing analysis, inverse design and constrained optimization computer codes. They should also try to implement methods for the acceleration of iterative algorithms in arbitrary systems of partial differential equations on highly clustered grids [266]. To make the entire design optimization software package as flexible as possible and responsive to the user's needs, it should be implemented on a cluster of disparate microcomputers, workstations, a vector multiprocessor, and a massively parallel processor.

Such a design tool should feature the combination of fully 3-D (not 2-D or quasi 3-D) aero and thermal capabilities dedicated to address user specified design objectives and improvements, and should provide:

- optimization of surface heat fluxes and temperatures for minimum coolant flow rate,
- minimization of the number of coolant flow passages,
- optimization of thicknesses of walls and interior struts, and
- determination of convective heat transfer coefficients on surfaces of 3-D coolant passages.

These design objectives could be accomplished using a number of conceptually different approaches. One specific set of scenarios that has been developed by the author can be summarized as follows.

13.5 Optimization of Surface Heat Fluxes and Temperatures for Minimum Coolant Flow Rate

As a by-product of the aerodynamic inverse design using Navier-Stokes computation of the hot gas flow field (with the specified temperatures as thermal boundary conditions on the hot surface) the corresponding hot surface heat flux distribution can be obtained. Since one of the global objectives is to minimize the coolant mass flow rate, the total integrated hot surface heat flux must be minimized. This can be achieved efficiently by utilizing a hybrid genetic evolution/gradient search constrained optimizer [267] and a reliable thermal boundary layer code. Input to such a code can be the hot surface temperature distribution. This temperature distribution can be discretized using β-splines [268] so that the locations of β-spline vertices (control points) can serve as the design variables. Each perturbation to the location of the β-spline vertices will create different hot surface temperature distribution. Wherever the computed hot surface local temperatures are larger than the maximum allowable temperature specified by the designer, they can be explicitly locally reduced to the maximum allowable temperature.

This temperature distribution and an already optimized hot surface pressure distribution can be used as inputs to the thermal boundary layer code. The hot surface heat flux distribution predicted by the thermal boundary layer code will be integrated to obtain the net heat input to the 3-D structure. After the hot surface temperatures have converged to their values that are compatible with the minimum net heat input to the structure, the aerodynamic shape inverse design Navier-Stokes code can be run again subject to these optimized hot surface temperatures. The resulting 3-D aerodynamic external shape will be slightly different than after the first inverse shape design and if necessary, the entire hot surface thermal optimization will be repeated with the redesigned external shape.

This repetitive simultaneous minimization of the integrated hot surface heat flux and the truncation and maximization of the hot surface local temperatures will converge to the final aerodynamically optimized 3-D external shape that satisfies optimized surface pressure distribution, maximum allowable surface temperature, and compatible hot surface thermal boundary conditions. The minimized integrated hot surface heat fluxes imply a minimized coolant flow rate requirement. Notice that this entire process does not require knowledge of the thermal field and the coolant flow passage configuration inside the internally cooled structure.

13.6 Minimization of the Number of Coolant Flow Passages

Thermal boundary condition inverse determination BEM codes can than use the hot surface temperatures and heat fluxes and non-iteratively predict the distribution of temperatures and heat fluxes on walls of the coolant passages. Convective heat transfer coefficients can then be computed on the walls of the coolant passages. These coefficients might locally exceed the realistic values attainable with the coolant flow in smooth passages. The locally varying heat convection coefficients should, therefore, be limited to their maximum allowable values. The corresponding modified heat fluxes on the walls of the coolant passages can then be determined. These "cold" heat fluxes and temperatures will then be submitted to the BEM inverse code that will determine the corresponding "hot" temperatures and heat fluxes. The resulting computed hot surface temperatures and heat fluxes will be different from the optimized hot surface values. The difference between the computed and the previously optimized hot surface temperatures and heat fluxes will then serve as the forcing function in a constrained optimization code. It will drive the sizes of unnecessary coolant passages to zero, while relocating, resizing, and reshaping the minimum necessary number of the passages until the differences in the computed and the optimized hot surface heat fluxes and temperatures are negligible. Minimum allowable distances among the coolant passages or from the thermal barrier coating interface will serve as constraints (Figure 86 and Figure 87).

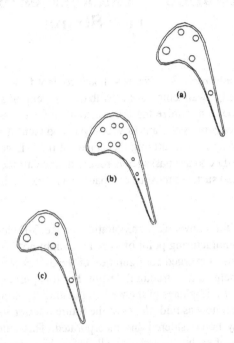

Figure 86   Minimization of the number of circular cross-section coolant passages inside
a ceramically coated turbine blade airfoil: a) target geometry; b) initial guess,
and c) an almost converged final result of the inverse shape design [245].

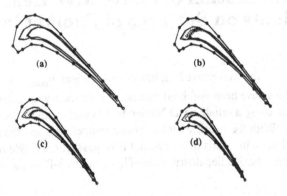

Figure 87   Geometric history of the optimization of a single coolant flow passage in a
three-dimensional turbine blade showing sections at (a) r = 0, (b) r = 0.25, (c)
r = 0.75, and (d) r = 1.0 subject to the specified temperatures and heat fluxes
on the blade hot surface and temperatures on the coolant passage wall [253],
[254].

13.7 Optimization of Thicknesses of Walls and Interior Struts

Next, the designer could run a 3-D Navier-Stokes coolant flow analysis code and compare its predicted coolant passage wall temperatures with the temperatures obtained in the previous task. The objective could be to minimize the difference in coolant passage surface temperatures obtained from the Navier-Stokes coolant flow analysis code and the heat conduction BEM code. This will be achieved by further altering the shapes of the 3-D coolant passages. This iterative modification of the 3-D coolant passages is required since all thermal boundary conditions on the external and internal surfaces must be compatible with each other and with the corresponding flow fields.

In addition, the converged configuration of the 3-D coolant passages might not be desirable from the manufacturing point of view and might not be structurally sound. Thus, the designer should use the converged configuration of the 3-D coolant passages only as an intelligent initial guess to help in the structural design. For this purpose the designer might wish to specify the locations of mid-planes of the walls separating the converged shapes of 3-D coolant passages and treat them now as mid-planes of the future interior struts. Other design constraints and requirements may be considered and incorporated. This automatically defines the initial thickness distributions of each strut and the wall. These thicknesses can be parameterized with a small number of β-spline coefficients which can then be the design variables in the optimization process.

13.8 Determination of Convective Heat Transfer Coefficients on Surfaces of Coolant Passages

Since the local values for the convective heat transfer coefficient, h_{conv}, on the surfaces of the 3-D coolant flow passages have been specified thus far, it is necessary to compute their actual values. This can be done using a reliable 3-D Navier-Stokes code with the best turbulence model available at the time. With the specified inlet coolant temperature and the already determined temperature distribution on the walls of the coolant flow passages, a single run with the Navier-Stokes code should predict a detailed distribution of h_{conv} on the 3-D surfaces of the coolant flow passages.

If the predicted values of h_{conv} are significantly different from those used in the previous tasks, the last two tasks should be repeated until convergence.

13.9 Conclusions

It is possible for thermal design and optimization to perform relatively efficiently, even in a fully 3-D case, if BEM codes are used for thermal field analysis and inverse determination of unknown boundary conditions and optimization is performed using a hybrid genetic evolution/gradient search constrained optimization algorithm. One specific conjugate heat transfer design and optimization scenario has been suggested for which all of the individual tasks have been proven to work by the author's research team. The main uncertainty of the entire conjugate heat transfer design process still rests with the issue of reliability of turbulence models in typical 3-D Navier-Stokes flow-field analysis codes to predict surface temperatures and heat fluxes.

13.10 References

[238] **Brebbia, C. A.**
The Boundary Element Method for Engineers, John Wiley & Sons, New York, 1978.

[239] **Brebbia, C. A., Dominguez, J.**
Boundary Elements, An Introductory Course, McGraw-Hill Book Company, New York, 1989.

[240] **Kennon, S. R., Dulikravich, G. S.**
The Inverse Design of Internally Cooled Turbine Blades, ASME Journal of Engineering for Gas Turbines and Power, Vol. 107, pp. 123-126, January 1985.

[241] **Kennon, S. R., Dulikravich, G. S.**
Inverse Design of Multiholed Internally Cooled Turbine Blades, International Journal of Numerical Methods in Engineering, Vol. 22, No. 2, pp. 363-375, 1986.

[242] **Kennon, S. R., Dulikravich, G. S.**
Inverse Design of Coolant Flow Passages Shapes With Partially Fixed Internal Geometries, International Journal of Turbo & Jet Engines, Vol. 3, No. 1, pp. 13-20, 1986.

[243] **Chiang, T. L., Dulikravich, G. S.**
Inverse Design of Composite Turbine Blade Circular Coolant Flow Passages, ASME Journal of Turbomachinery, Vol. 108, pp. 275-282, Oct. 1986.

[244] **Dulikravich, G. S.**
Inverse Design and Active Control Concepts in Strong Unsteady Heat Conduction, Applied Mechanics Reviews, Vol. 41, No. 6, June 1988, pp. 270-277.

[245] **Dulikravich, G. S., Kosovic, B.**
Minimization of the Number of Cooling Holes in Internally Cooled Turbine Blades, ASME paper 91-GT-103, ASME Gas Turbine Conf., Orlando, Florida, June 2-6, 1991; *also in* Internat. Journal of Turbo & Jet Engines, Vol. 9. No. 4, pp. 277-283, 1992.

[246] **Dulikravich, G. S.**
Inverse Design of Proper Number, Shapes, Sizes and Locations of Coolant Flow Passages, in Proceedings of the 10th Annual CFD Workshop (ed. R. Williams), NASA MSFC, Huntsville, AL, April 28-30, 1992, NASA CP-3163, Part 1, pp. 467-486, 1992.

[247] **Dulikravich, G. S., Martin, T. J.**
Determination of Void Shapes, Sizes and Locations Inside an Object With Known Surface Temperatures and Heat Fluxes, in Proceedings of the IUTAM Symposium on Inverse Problems in Engineering Mechanics (editors: M. Tanaka and H.D. Bui), Tokyo, Japan, May 11-15, 1992; also in Springer-Verlag, pp. 489-496, 1993.

[248] **Dulikravich, G. S., Martin, T. J.**
Determination of the Proper Number, Locations, Sizes and Shapes of Superelliptic Coolant Flow Passages in Turbine Blades, in Proceedings of the International Symposium on Heat and Mass Transfer in Turbomachinery (ICHMT) (ed. R.J. Goldstein, A. Leontiev, and D. Metzger), Athens, Greece, August 24-28, 1992.

[249] **Dulikravich, G. S., Martin, T. J.**
Design of Proper Super-Elliptic Coolant Passages in Coated Turbine Blades With Specified Temperatures and Heat Fluxes, AIAA paper 92-4714, 4th AIAA/AHS/ASEE Symposium on Multidisciplinary Analysis & Optimization, Cleveland, Ohio, Sept. 21-23,1992; also in AIAA Journal of Thermophysics and Heat Transfer, Vol. 8, No. 2, pp. 288-294, April - June 1994.

[250] **Dulikravich, G. S., Martin, T. J.**
Inverse Design of Super-Elliptic Cooling Passages in Coated Turbine Blade Airfoils, AIAA Journal of Thermophysics and Heat Transfer, Vol. 8, No. 2, April-June, 1994, pp. 288-294.

[251] **Dulikravich, G. S., Chiang, T. L., Hayes, L. J.**
Inverse Design of Coolant Flow Passages in Ceramically Coated Scram-Jet Combustor Struts, ASME WAM'86, Anaheim, CA, December 1986, in Proceedings of Symposium on Numerical Methods in Heat Transfer (editors: M. Chen and K. Vafai), ASME HTD-Vol. 62, pp. 1-6, 1986.

[252] **Martin, T. J., Dulikravich, G. S.**
Inverse Design of Threedimensional Shapes With Overspecified Thermal Boundary Conditions, Monograph on Inverse Problems in Mechanics, (editor: S. Kubo), Atlanta Technology Publications, Atlanta, GA, September 1993, pp. 128-140.

[253] **Dulikravich, G. S., Martin, T. J.**
Three-Dimensional Coolant Passage Design for Specified Temperatures and Heat Fluxes, AIAA paper 94-0348, AIAA Aerospace Sciences Meeting, Reno, NV, January 10-13, 1994.

[254] **Dulikravich, G. S., Martin, T. J.**
Geometrical Inverse Problems in Three-Dimensional Non-Linear Steady Heat Conduction, Engineering Analysis with Boundary Elements, Vol. 15, 1995, pp. 161-169.

[255] **Dulikravich, G. S., Martin, T. J.**

Inverse Shape and Boundary Condition Problems and Optimization in Heat Conduction, Chapter 10 in Advances in Numerical Heat Transfer, Vol. 1, (editors: W. J. Minkowycz and E. M. Sparrow), Taylor and Francis, 1996.

[256] **Martin, T. J., Dulikravich, G. S.**
A Direct Approach to Finding Unknown Boundary Conditions in Steady Heat Conduction, Proc. of 5th Annual Thermal and Fluids Workshop, NASA CP-10122, NASA LeRC, Ohio, Aug. 16-20, 1993, pp. 137-149.

[257] **Martin, T. J., Dulikravich, G. S.**
Inverse Determination of Temperatures and Heat Fluxes on Inaccessible Surfaces, Boundary Element Technology IX, Computational Mechanics Publications, Southampton (editors: C. A. Brebbia and A. Kassab), pp. 69-76, 1994.

[258] **Dulikravich, G. S., Martin, T. J.**
Inverse Problems and Design in Heat Conduction, in Proceedings of 2nd IUTAM International Symposium on Inverse Problems in Engineering Mechanics (editors: H.D. Bui, M. Tanaka, M. Bonnet, H. Maigre, E. Luzzato, and M. Reynier), Paris, France, November 2-4, 1994, A. A. Balkema, Rotterdam, pp. 13-20, 1994.

[259] **Martin, T. J., Dulikravich, G. S.**
Finding Unknown Surface Temperatures and Heat Fluxes in Steady Heat Conduction, in Proceedings of 4th Intersociety Conference on Thermal Phenomena in Electronic Systems (editors: A. Ortega and D. Agonafer), Washington, D. C., May 4-7, 1994, pp. 214-221; also in IEEE Transactions on Components, Packaging and Manufacturing Technology (CPMT) - Part A, Vol. 18, No. 3, Sept. 1995, pp. 540-545.

[260] **Martin, T. J., Dulikravich, G. S.**
Inverse Determination of Boundary Conditions in Steady Heat Conduction With Heat Generation, in Symposiums on Conjugate Heat Transfer, Inverse Problems, and Optimization, and Inverse Problems in Heat Transfer, (editors: W. J. Bryan and J. V. Beck), ASME National Heat Transfer Conf., Portland, OR, Aug. 6-8,1995, ASME HTD-Vol. 312, pp. 39-46.

[261] **Martin, T. J., Dulikravich, G. S.**
Finding Temperatures and Heat Fluxes on Inaccessible Surfaces in 3-D Coated Rocket Nozzles, in Proceedings of 1995 JANNAF (Joint Army-Navy-NASA-Air Force) Propulsion and Subcommittee Joint Meeting, Tampa, FL, December 4-8, 1995.

[262] **Martin, T. J., Dulikravich, G. S.**
Inverse Determination of Boundary Conditions and Sources in Steady Heat Conduction With Heat Generation, ASME Journal of Heat Transfer, Vol. 110, No. 3, Aug. 1996.pp. 546-554.

[263] **Martin, T. J., Dulikravich, G. S.**
Determination of Temperatures and Heat Fluxes on Surfaces of Multidomain Three-Dimensional Electronic Components (with T. J. Martin), Symposium on Application of CAE/CAD to Electronic Systems, 1996 ASME Int. Mechanical Eng. Congress and Expo., Atlanta, GA, November 17-22, 1996.

[264] **Martin, T. J., Dulikravich, G. S.**
Inverse Determination of Temperatures and Heat Fluxes on Surfaces of 3-D Objects, PanAmerican Congress of Applied Mechanics (PACAM-V), San Juan, Puerto Rico, Jan. 2-4, 1997.

[265] **Press, W. H., Teukolsky, S. A., Vetterling, W. T., Flannery, B. P.**
Numerical Recipes in FORTRAN: The Art of Scientific Computing, Second Edition, Cambridge University Press, 1992.

[266] **Choi, K. Y., Dulikravich, G. S.**
Acceleration of Iterative Algorithms on Highly Clustered Grids, AIAA Journal, Vol. 34, No. 4, April 1996, pp. 691-699.

[267] **Foster, N. F., Dulikravich, G. S., Bowles, J.**
Three-Dimensional Aerodynamic Shape Optimization Using Genetic Evolution and Gradient Search Algorithms, AIAA paper 96-0555, AIAA Aerospace Sciences Meeting, Reno, NV, January 15-19, 1996.

[268] **Farin, G.**
Curves and Surfaces for Computer Aided Geometric Design, Second Edition, Academic Press, Boston, 1990.

CHAPTER 14

STRUCTURAL INVERSE DESIGN AND OPTIMIZATION

G.S. Dulikravich

The Pennsylvania State University, University Park, PA, USA

14.1 Introduction

Design of structural components for specified aerodynamic and dynamic loads can be achieved with an extensive use of a reliable and versatile stress-deformation prediction code and a constrained optimization code. The finite element method (FEM) is the favorite method for structural analysis because of its adaptability to complex 3-D structural configurations and the ability to easily account for a point-by-point variation of physical properties of the material. With the recent improvements in the sparse matrix solver algorithms, the finite element techniques are also becoming competitive with the finite difference techniques in terms of the computer memory and computing time requirements. For relatively smaller problems in elasticity, it is even more advantageous to use boundary element method (BEM) because it is faster and more reliable since it is non-iterative and it requires only surface discretization.

14.2 Optimization in Elasticity

The field of structural optimization [269]-[273] has a considerably longer history than aerodynamic shape optimization. Almost every textbook on optimization involves examples from linear elastostatics since these types of problems usually result in smooth convex function spaces that have continuous and finite sensitivity derivatives, making them ideal for gradient search optimization. As the geometric complexity and the diversity of materials involved in aerospace structures has increased, so has the demand for a more robust non-gradient search optimization algorithms. Consequently, the majority of the present-day structural optimization is performed using different variations of genetic evolution search strategy and a hybrid gradient/genetic approach [274], [275]. This is quite evident when researching the literature dealing with structures made of composite materials, ceramically coated structures, and smart structures. In the case of smart structures, their "smart" attribute comes from the ability of such materials to respond with a desired degree of deformation to the applied pressure, thermal, electric or magnetic field. This automatic response can be very fast and can be used for active control of the structural shape in the regions exposed to the air flow. This influences the flow field, surface heat transfer and the aerodynamic forces acting on the structure.

Structural optimization is routinely used for the purpose of achieving aeroelastic tailoring. Since this requires specifications of a large number of constraints in the form of desired local deformations, this application can also be qualified as a *de facto* inverse structural shape design. The objective is to find the appropriate shape and orientation of each layer of a composite material to be formed so that the final object (for example, a helicopter rotor blade, an airplane wing, etc.) will have the most uniform stress distribution (thus minimum weight) and, when loaded, will deform into a desired form.

14.3 Inverse Problems in Elastostatics

An elastostatic problem is well-posed when the geometry of the general 3-D multiply-connected object is known and either displacement vectors, **u**, or surface traction vectors, **p**, are specified everywhere on the surface of the object. The elastostatic problem becomes ill-posed when either: a) a part of the object's geometry is not known, or b) when both u and p are unknown on certain parts of the surface. Both types of inverse problems can be solved only if additional information is provided. This information should be in the form of over-specified boundary conditions where both u and p are simultaneously provided at least on certain surfaces of the 3-D body.

The inverse determination of locations, sizes and shapes of unknown interior voids subject to overspecified stress-strain outer surface field is a common inverse design problem in elasticity [276]-[279]. The general approach is to formulate a cost function that measures a sum of least squares differences in the surface values of given and computed stresses or deformations for a guessed configuration of voids. This cost function is then minimized using any of the standard optimization algorithms by perturbing the number, sizes, shapes and locations of the guessed voids. Thus, the process is identical to the already described inverse design of coolant flow passages subject to over-specified surface thermal conditions [280]. It should be pointed out that this approach to inverse design of interior cavities and voids can generate interior configurations that are potentially non-unique.

14.4 Inverse Determination of Elastostatic Boundary Conditions

Another type of inverse problem in elastostatics is to deduce displacements and tractions on surfaces where such information is unknown or inaccessible, although the geometry of the entire 3-D configuration is given. It is often difficult and even impossible to place strain gauges and take measurements on a particular surface of a solid body either due to its small size or geometric inaccessibility or because of the severity of the environment on that surface. With our inverse method [281] these unknown elastostatics boundary values can be deduced from additional displacement and surface traction measurements made at a finite number of points within the solid or on some other surfaces of the solid. This approach is robust and fast since it is non-iterative. A similar inverse boundary value formulation has been shown [282] to compute meaningful and accurate thermal fields during a single analysis using a straight-forward modification to the BEM non-linear heat conduction analysis code. An example of the concept follows using the BEM.

In general elastostatics, we can write for any discretized boundary point "i" and for each direction "l" a boundary integral equation [283]

$$c_1^i \cdot u_1^i + \int_\Gamma u_k p_{1k}^* d\Gamma = \int_\Gamma p_k u_{1k}^* d\Gamma \tag{90}$$

where the asterisk designates fundamental solutions and the term c_l is obtained with some special treatment of the surface integral on the left hand side [283]. Explicit calculation of this value can be obtained by augmenting the surface integral over the singularity that occurs when the integral includes the point "i". Fortunately, explicit calculation is not necessary as it can be obtained using the rigid body motions. The boundary G is discretized into N_{sp} surface panels connected between N nodes. The functions **u** and **p** are quadratically distributed over each panel with adjacent panels sharing nodes such that there will be twice as many boundary nodes as there are surface panels. A transformation from the global (x,y) coordinate system to a localized boundary fitted (ξ, η) coordinate system is required in order to numerically integrate each surface integral using Gaussian quadrature. The displacements and tractions are defined in terms of three nodal values and three quadratic interpolation functions. The whole set of boundary integral equations can be written in matrix form (ommiting the body forces for simplicity) as

$$[H]\{U\} = [G]\{P\} \tag{91}$$

where the vectors $\{U\}$ and $\{P\}$ contain the nodal values of the displacement and traction vectors. Each entry in the [H] and [G] matrices is developed by properly summing the contributions from each numerically integrated surface integral. The surface tractions were allowed to be discontinuous between each neighboring surface panel to allow for proper corner treatment. The set of boundary integral equations will contain a total of 2N equations and 6N nodal values of displacements and surface tractions.

For a well-posed boundary value problem, at least one of the functions, **u** or **p**, will be known at each boundary node (either Dirichlet or von Neumann boundary condition) so that the equation set will be composed of 2N unknowns and 2N equations. Since there are two distinct traction vectors at corner nodes, the boundary conditions applied there should include either two tractions or one displacement and one traction. If only displacements are specified across a corner node, special treatment is required [283].

For an ill-posed boundary value problem, both **u** and **p** should be enforced simultaneously at certain boundary nodes, while either **u** or **p** should be enforced at some of the other boundary nodes, and nothing enforced at the remaining boundary nodes. For the simple example of a quadrilateral plate with four nodes, if at two boundary nodes both **u** = **U** and **p** = **P** are known, but at the other two nodes neither **u** nor **p** is known, the BEM equation set before any rearrangement appears as

$$
\begin{bmatrix} h_{11} & h_{12} & h_{13} & h_{14} \\ h_{21} & h_{22} & h_{23} & h_{24} \\ h_{31} & h_{32} & h_{33} & h_{34} \\ h_{41} & h_{42} & h_{43} & h_{44} \end{bmatrix} \cdot \begin{bmatrix} U_1 \\ u_2 \\ U_3 \\ u_4 \end{bmatrix} = \begin{bmatrix} g_{11} & g_{12} & g & g_{14} \\ g_{21} & g_{22} & g_{23} & g_{24} \\ g_{31} & g_{32} & g_{33} & g_{34} \\ g_{41} & g_{42} & g_{43} & g_{44} \end{bmatrix} \cdot \begin{bmatrix} P_1 \\ P_2 \\ P_3 \\ P_4 \end{bmatrix}
\tag{92}
$$

where each of the entries in the [H] and [G] matrices is a 3x3 submatrix in 3-D elastostatics. Straight-forward algebraic manipulation yields the following set

$$
\begin{bmatrix} h_{11} & -g_{12} & h_{14} & -g_{14} \\ h_{22} & -g_{22} & h_{24} & -g_{24} \\ h_{32} & -g_{32} & h_{34} & -g_{34} \\ h_{42} & -g_{42} & h_{44} & -g_{44} \end{bmatrix} \cdot \begin{bmatrix} u_2 \\ P_2 \\ u_4 \\ P_4 \end{bmatrix} = \begin{bmatrix} -h_{11} & g_{11} & -h_{13} & g_{13} \\ -h_{21} & g_{21} & -h_{23} & g_{23} \\ -h_{31} & g_{31} & -h_{33} & g_{33} \\ -h_{41} & g_{41} & -h_{43} & g_{43} \end{bmatrix} \cdot \begin{bmatrix} U_1 \\ P_1 \\ P_3 \\ P_3 \end{bmatrix} = \begin{bmatrix} f_1 \\ f_2 \\ f_3 \\ f_4 \end{bmatrix}
\tag{93}
$$

The right-hand side vector {**F**} is known and the left-hand side remains in the form [A]{**X**}. Once the matrix [A] is solved, the entire **u** and **p** fields within the solid can be easily deduced from the integral formulation. The equation set [A]{**X**} = {**F**} resulting from our inverse boundary value formulation is highly singular and most standard matrix solvers will produce an incorrect solution. Singular Value Decomposition (SVD) methods [284] can be used to solve such problems accurately. The number of unknowns in the equation set need not be the same as the number of equations [284], so that virtually any combination of boundary conditions will yield at least some solution. Additional equations may be added to the equation set if **u** measurements are known at locations within the solid in order to enhance the accuracy of the inverse steady boundary condition algorithm. A proper physical solution will be obtained if the number of equations equals or exceeds the number of unknowns. If the number of equations is less than the number of unknowns, the SVD method will find one solution, although it does not necessarily have to be the proper solution from the physical point of view [285]. Thus, the more overspecified data is made available, the more accurate and unique the predicted boundary values will be.

The accuracy of the inverse boundary condition code was verified [281] on simple 2-D geometry consisting of an infinitely long thick-walled pipe subject to an internal gauge pressure. The shear modulus for this problem was $G = 8.0 \times 10^4$ N/mm^2 and Poisson's ratio was $\nu = 0.25$. The radius of the inner surface of the pipe was 10 mm and the outer radius was 25 mm. The inner and outer boundaries were discretized with 12 quadratic panels each. The internal gauge pressure was specified to be $P_r = 100$ N/mm^2, while the outer boundary was specified with a zero surface traction. Figure 88 depicts a contour plot of constant values of stress tensor components σ_{yy} that was computed using the analysis version of our second-order accurate BEM elastostatic code. The numerical results of this well-posed boundary value problem were then used as boundary conditions applied to the following two ill-posed problems.

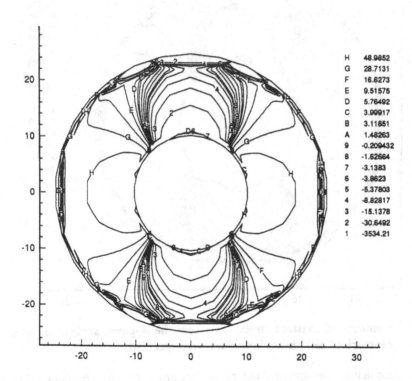

H	48.9852
G	28.7131
F	16.6273
E	9.51575
D	5.76492
C	3.99917
B	3.11851
A	1.48263
9	-0.209432
8	-1.62664
7	-3.1383
6	-3.8623
5	-5.37803
4	-8.82817
3	-15.1378
2	-30.6492
1	-3534.21

Figure 88 **Contours of constant stress, σ_{yy}, from the wellposed analysis of an annular pressurized disk.**

First, the displacement vectors computed on the inner circular boundary were applied as over-specified boundary conditions in addition to the surface tractions already enforced there. At the same time nothing was specified on the outer circular boundary. Figure 89 represents the contour plot of σ_{yy} that was obtained with our inverse boundary value BEM code. This stress averaged a much larger error, about 3.0%, with some asymmetry in the stress field, when compared with the analysis results.

Next, the displacement vectors computed on the outer circular boundary by the well-posed numerical analysis were used to over-specify the outer circular boundary. At the same time nothing was specified on the inner circular boundary. Figure 90 the contour plot of σ_{yy} as computed by the inverse BEM technique. The predicted inner surface deformations were in error by less than 0.1%, while the predicted inner surface stresses averaged less than a 1.0% error as compared to the analysis results.

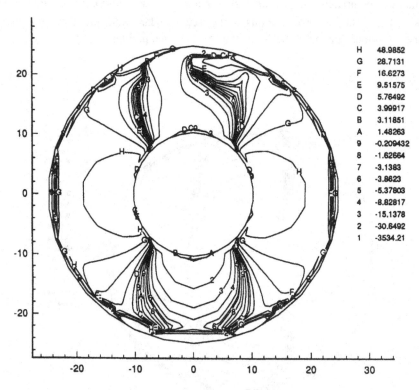

H	48.9852
G	28.7131
F	16.6273
E	9.51575
D	5.76492
C	3.99917
B	3.11851
A	1.48263
9	-0.209432
8	-1.62664
7	-3.1383
6	-3.8623
5	-5.37803
4	-8.82817
3	-15.1378
2	-30.6492
1	-3534.21

Figure 89   **Contours of constant stress, σ_{yy}, from the ill-posed analysis of an annular pressurized disk: inner boundary over-specified.**

It seems that an over-specified outer boundary produces a more accurate solution than one having an over-specified inner boundary. It was also shown that as the over-specified boundary area or the resolution in the applied boundary conditions was decreased, the amount of over-specified data also decreases, and thus the accuracy of the inverse boundary value technique deteriorates.

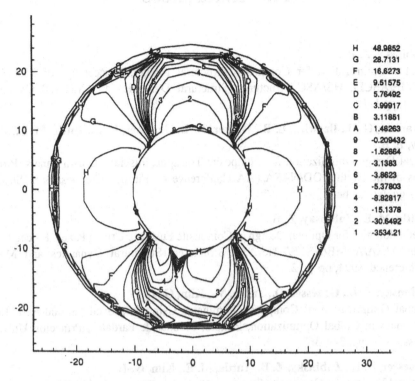

H	48.9852
G	28.7131
F	16.6273
E	9.51575
D	5.76492
C	3.99917
B	3.11851
A	1.48263
9	-0.209432
8	-1.62664
7	-3.1383
6	-3.8623
5	-5.37803
4	-8.82817
3	-15.1378
2	-30.6492
1	-3534.21

Figure 90   **Contours of constant stress, σ_{yy}, from the ill-posed analysis of an annular pressurized disk: outer boundary over-specified.**

14.5 Conclusions

Structural design of complex 3-D composite structures is possible only by using reliable optimization codes. For simpler configurations with smaller numbers of variables it is also possible to use inverse shape design methodologies based on over-specified surface tractions and deformations. Inverse shape design and inverse boundary condition determination can be performed using both BEM and finite element techniques. Hybrid optimization algorithms [286] that are based on genetic evolution, simulated annealing, fuzzy logic and neural networks and that involve manufacturing constraints [287]-[288] will be the candidates for active elastodynamic/aeroelastic control of smart structures.

14.6 References

[269] **Zabinsky, Z. B.**
Global Optimization for Composite Structural Design, Proceedings of 35th AIAA/
ASME/ASCE/AHS/ASC Structures, Structural Dynamics and Material Conference,
1994, pp 1-7.

[270] **Swanson, G. D., Ilcewicz, L. B., Walker, T. H., Graesser, D. L., Tuttle, M. E., Zabinsky, Z. B.**
Local Design Optimization for Composite Transport Fuselage Crown Panels, Proceedings of the Ninth DOD/NASA/FAA Conference on Fibrous Composites in Structural
Design, November 1991, pp 1-20.

[271] **Tuttle, M. E., Zabinsky, Z. B.**
Methodologies for Optimal Design of Composite Fuselage Crown Panels Proceedings of
35th AIAA/ASME/ASCE/AHS/ASC Structures, Structural Dynamics and Material
Conference, 1994, pp 1-12.

[272] **Zabinsky, Z. B., Graesser, D., Tuttle, M., Kim, G.-I.**
Global Optimization of Composite Laminates Using Improved Hit-and-Run, Recent
Advances in Global Optimization, Editors: Floudas & Pardalos, Princeton University
Press, 1992, pp. 344-367.

[273] **Graesser, D. L., Zabinsky, Z. B., Tuttle, M. E., Kim, G.-I.**
Optimal Design of a Composite Structure, Composite Structures, Vol. 24, 1993, pp. 273-
281.

[274] **Graesser, D. L., Zabinsky, Z. B., Tuttle, M. E., Kim, G.-I.**
Designing Laminated Composites Using Random Search Techniques, Composite Structures, Vol. 18, 1991, pp. 311-325.

[275] **Sudipto, N., Zabinsky, Z. B., Tuttle, M. E.**
Optimal Design of Composites Using Mixed Discrete and Continuous Variables,
Processing, Design and Performance of Composite Materials, ASME MD-Vol. 52, 1994,
pp. 91-107

[276] **Bezzera, L. M., Saigal, S.**
A boundary element formulation for the inverse elastostatics problem (IESP) of flaw
detection, International Journal for Numerical Methods in Engineering, Vol. 36, 1993,
pp. 2189 -2202.

[277] **Kassab, A. J., Moslehy, F. A., Daryapurkar, A.**
Detection of cavities by inverse elastostatics boundary element method: experimental
results, in Boundary Element Technology IX, Computational Mechanics Publications,
Southampton, Editors: C. A. Brebbia and A. Kassab, 1994, pp. 85-92.

[278] **Tanaka, M., Yamagiwa, K.**
A Boundary Element Method for Some Inverse Problems in Elastodynamics, Appl. Math. Modell., Vol. 13, 1989, pp. 307-312.

[279] **Zabaras, N., Morellas, V., Schnur, D.**
A Spatially Regularized Solution of Inverse Elasticity Problems Using the Boundary Element Method, Communications in Appl. Numer. Meth., Vol. 5, 1989, pp. 547-553.

[280] **Dulikravich, G. S., Martin, T. J.**
Inverse design of super-elliptic cooling passages in coated turbine blade airfoils, AIAA Journal of Thermophysics and Heat Transfer, 8, No. 2, 288-294, April-June, 1994.

[281] **Martin, T. J., Halderman, J. D., Dulikravich, G. S.**
An Inverse Method for Finding Unknown Surface Tractions and Deformations in Elastostatics, Computers and Structures, Vol. 56, No. 5, Sept. 1995, pp. 825-836.

[282] **Martin, T. J., Dulikravich, G. S.**
Inverse Determination of Boundary Conditions in Steady Heat Conduction with Heat Generation, ASME Journal of Heat Transfer, Vol. 118, No. 3, August 1996, pp. 546-554.

[283] **Brebbia, C. A.**
The Boundary Element Method for Engineers, John Wiley & Sons, New York, 1978.

[284] **Press, W. H., Teukolsky, S. A., Vetterling, W. T., Flannery, B. P.**
Numerical Recipes in FORTRAN, Second Edition, Cambridge University Press, 1992.

[285] **Okuma, M., Kukil, S.**
Correction of Finite Element Models Using Experimental Modal Data for Vibration Analysis, Monograph on Inverse Problems in Mechanics, Editor: S. Kubo, Atlanta Technology Publications, Atlanta, GA, 204-211, September, 1993.

[286] **Zabinsky, Z. B., Smith, R., McDonald, J., Romeijn, H., Kaufman, D.**
Improving Hit-and-Run for Global Optimization, Journal of Global Optimization, Vol. 3, 1993, pp. 171-192.

[287] **Kristinsdottir, B., Zabinsky, Z. B., Tuttle, M. E., Csendes, T.**
Incorporating Manufacturing Tolerances in Near-Optimal Design of Composite Structures, Engineering Optimization, 1996, Vol. 26, pp. 1-23.

[288] **Kristinsdottir, B. P., Zabinsky, Z. B.**
Including Manufacturing Tolerances in Composite Design, Proceedings of 35th AIAA/ASME/ASCE/AHS/ASC Structures, Structural Dynamics and Material Conference, 1994, pp 1-10.

CHAPTER 15

MULTIDISCIPLINARY INVERSE DESIGN AND OPTIMIZATION (MIDO)

G.S. Dulikravich

The Pennsylvania State University, University Park, PA, USA

15.1 Introduction

Designing for improved performance and life expectancy of high speed transport configurations is traditionally conducted by performing a repetitive sequence of uncoupled, single-discipline analyses involving flow field, temperature field, stress-strain field, structural dynamics, manufacturability tradeoffs and a large amount of personal designer's experience and intuition [289]-[295]. Since the entire aircraft system is seemingly highly coupled, it would be plausible that both analysis and design should be performed using an entirely new generation of computer codes that solve a huge system of partial differential equations governing aerodynamics, elastodynamics, heat transfer inside the structure, dynamics, manufacturing cost estimates, etc. simultaneously. This approach offers very stable computation since all boundary and interfacing conditions are incorporated implicitly. On the other hand, this approach might not be the most computationally economical since different subsystems (Navier-Stokes equations, elastodynamic equations, heat conduction equation, Maxwell's equations, etc.) that form such a complex mathematical system have vastly different eigenvalues and consequently converge at significantly different rates to a steady state solution. In addition, a rigorous analysis can show that even seemingly highly coupled systems are only loosely coupled and can be analyzed semi-sequentially [296]. Such a semi-sequential approach is presently used by most researchers and the industry since it can utilize most of the existing analysis and inverse design and optimization software as ready and interchangeable modules with minimum time invested in their modifications. Nevertheless, this approach is much more prone to global instability because of the often unknown and inadequately treated boundary and interface conditions.

In the remaining part of this article the focus will be on the computational grid, acceleration of iterative algorithms and parallelization and networking issues that are pertinent to the MIDO efforts.

15.2 Computational Grid Generation

It is well understood that automatic discretization (boundary-conforming computational grid generation) is the main bottleneck of the entire computational aerodynamics. Typically, it takes more time to generate an acceptable new grid for a new realistic 3-D aerodynamic configuration than it takes to predict the 3-D flow field around it. Since any shape inverse design and optimization effort implies repetitive generation of the 3-D grid, it is quite clear that developing a user-friendly, fast and reliable 3-D computational grid generation code is an extremely important issue. The existing automatic 3-D grid generation codes accept surface grid coordinates as an input and then generate the coordinates of field grid points. This approach is not reliable in the sense that there is no guarantee that some of the generated grid cells will not fold-over, or become needle-like, or that excessively large cells will not neighbor excessively-small grid cells, or that the grid will not be sufficiently clustered in the regions of interest, or that the grid will not become excessively non-orthogonal. Such problems cause significantly slower convergence of the analysis iterative algorithms, they create significantly larger computational errors, cause numerical instability, and often lead to outright divergence of the iterative process.

Any grid generation algorithm must be computationally efficient in terms of storage and execution time and able to accept even incomplete initial grids, that is, the grids that have only surface grid points resulting from a solid modeling software package (for example, CATIA). One technique capable of generating reliable 3-D grids in *an a posteriori* fashion is the grid optimization method which has been applied to 2-D [297] and 3-D [298] structured and block-structured [299] and to 2-D triangular unstructured grids including automatic solution-adaptive clustering [300]. This fully 3-D robust computational grid post-processing algorithm checks the entire original grid for possible negative Jacobians (existence of fold-over grid cells), untangles the grid lines, and optimizes the grid in a sense that it becomes maximally locally orthogonal and smoothly clustered in the regions of interest.

Notice that this grid generation procedure is especially suitable to the tightly coupled approached mentioned above. In this case, the grid from the flow field smoothly blends with the grid in the solid structure and the interior of the aircraft while clustering at the interface surfaces.

15.3 Convergence Acceleration and Reliability Enhancement

At present, it is possible to solve an impressive array of complex fluid dynamic problems by numerical techniques, but in many instances a minor change in the problem being computed can require an unwarranted amount of operator intervention, and/or an unexpectedly large increase in computer time. For example, a minor change in geometry or grid spacing sometimes has a major effect on convergence and can even lead to divergence, while coaxing a new problem through to a successfully converged solution can take many hours for even an experienced numerical analyst. Difficulties of this nature frequently dominate the cost of completing numerical solutions and can prevent the effective implementation of numerical techniques in design procedures.

Reduction of total computing time required by iterative algorithms for numerical integration of Navier-Stokes equations for 3-D, compressible, turbulent flows with heat transfer is an important aspect of making the existing and future analysis and inverse design codes widely acceptable as main components of optimization design tools. Reliability of analysis and design codes is an equally important item especially when varying input parameters over a wide range of values. Although a variety of methods have been tried, it remains one of the most challenging tasks to develop and extensively verify new concepts that will guarantee substantial reduction of computing time over a wide range of grid qualities (clustering, skewness, etc.), flow field parameters (Mach numbers, Reynolds numbers, etc.), types and sizes of systems of partial differential equations (elliptic, parabolic, hyperbolic, etc.). While a number of methods are capable of reducing the total number of iterations required to reach the converged solution, they require more time per iteration so that the effective reduction in the total computing time is often negligible.

The existing techniques for convergence acceleration are known to have certain drawbacks. Specifically, residual smoothing [301], although simple to implement, is a highly unreliable method, because it can offer either substantial reduction of number of iterations or it can abruptly diverge due to a poor choice of smoothing parameters [302]. Enthalpy damping [301] assumes constant total enthalpy which is incompatible with viscous flows including heat transfer. Multigridding in 3-D space is only marginally stable [303] when applied to non-smooth and non-orthogonal grids. GMRES method [304] based on conjugate gradients requires a large number of solutions to be stored per each cycle which is intractable in 3-D viscous flow computations. Power method [305] which is practically identical to the GNLMR method [306] works well with a multigrid code. Without multigridding, it is highly questionable if the power method would offer any acceleration when applied to a system of nonlinear partial differential equations. Despite the multigrid method's superior capability of effectively reducing low and high frequency errors yielding impressive convergence rates, its efficiency is significantly reduced on a highly-clustered grid [302], [303] and it is difficult to implement reliably. The preconditioning methods [307], although very powerful in alleviating the slow convergence associated with a stiff system for solving the low Mach number compressible flow equations, have not been shown to perform universally well on highly- clustered grids. Distributed Minimal Residual

(DMR) method [308], [309] in its present form has the same problem in transonic range. Numerous other methods have been published that are considerably more complex, while less reliable and effective.

The main commonality to all of these methods is that they all experience loss of their ability to reduce the computing time on highly-clustered non-orthogonal grids [307], [310] that are unavoidable for 3-D aerodynamic configurations and high Reynolds number turbulent flows.

Existing iterative algorithms are based on evaluating a correction, $\Delta \mathbf{Q}^t$, to each of the variables and then adding a certain fraction, $\omega \Delta \mathbf{Q}^t$, of the correction to the present value of the vector of variables, $\mathbf{Q}^t$; thus forming the next iterative estimates as

$$Q^{t+1} = Q^t + \omega \Delta Q^t \tag{94}$$

The optimal value of the relaxation factor, ω, can be determined from the condition that the future residual is minimized with respect to ω.

Instead, corrections from N consecutive iterations could be saved and added to the present value of the variable in a weighted fashion, where each of the consecutive iterative corrections is weighted by its own relaxation factor. This is the General Nonlinear Minimal Residual (GNLMR) acceleration method [306], summarized as

$$Q^{t+1} = Q^t + \omega^t \Delta Q^t + \omega^{t-1} \Delta Q^{t-1} + \dots + \omega^{t-N-1} \Delta Q^{t-N-1} \tag{95}$$

The optimal values of the relaxation factors, ω, can be determined from the condition that the future residual is minimized simultaneously with respect to each of the N values of ω's.

Now, let us consider an arbitrary system of M partial differential equations. If the GNLMR method is applied to each of the M equations so that each equation has its own sequence of N relaxation factors premultiplying its own N consecutive corrections, this defines the Distributed Minimal Residual (DMR) acceleration method [308], [309] formulated as

$$Q_1^{t+1} = Q_1^t + \omega_1^t \Delta Q_1^t + \omega_1^{t-1} \Delta Q_1^{t-1} + \dots + \omega_1^{t-N-1} \Delta Q_1^{t-N-1} \tag{96}$$

$$Q_M^{t+1} = Q_M^t + \omega_M^t \Delta Q_M^t + \omega_M^{t-1} \Delta Q_M^{t-1} + \dots + \omega_M^{t-N-1} \Delta Q_{1M}^{t-N-1} \tag{97}$$

The DMR is applied periodically where the number of iterations performed with the basic non-accelerated algorithm between two consecutive applications of the DMR is an input parameter. Here, ω's are the iterative relaxation parameters (weight factors) to be calculated and optimized, ΔQ's are the corrections computed with the non-accelerated iteration scheme, N denotes the total number of consecutive iteration steps combined when evaluating the optimum ω's, and M stands for the total number of equations in the system that is being iteratively solved.

The DMR method calculates optimum ω's to minimize the L-2 norm of the future residual of the system integrated over the entire domain, D. The present formulation of the DMR uses the same values of the N x M optimized relaxation parameters at every grid point, although different parts of the flow field converge at different rates.

An arbitrary system of partial differential equations governing an unsteady process can be written as $R = \dfrac{\partial Q}{\partial t} = L(Q)$, where Q is the vector of solution variables, t is the physical time, L is the differential operator and R is the residual vector.

Sensitivity-Based Minimum Residual (SBMR) method [311], [312] uses the fact that the future residual at a grid point depends upon the changes in Q at the neighboring grid points used in the local finite difference approximation. The sensitivities are determined by taking partial derivatives of the finite difference approximation of R_r (r=1,...,rmax where rmax is the number of equations in the system) with respect to each component of Q_{ms} (m = 1,..., M where M is the number of unknowns; s = 1,...,S where S is the number of surrounding grid points directly involved in the local discretization scheme). This information is then utilized to effectively extrapolate Q so as to minimize the future residual, R. Nine grid points located at (i-1,j-1; i-1,j; i-1, j+1; i,j-1; i,j; i,j+1; i+1,j-1; i+1,j; i+1,j+1) are used to formulate the global SBMR method for a two-dimensional problem when using central differencing compared to nineteen grid points for a three-dimensional case when using central differencing. This approach is different from the DMR [308], [309] method where the analytical form of R was differentiated.

Suppose that we are performing an iterative solution of an arbitrary evolutionary system using an arbitrary iteration algorithm. Suppose we know the solution vectors Q^t and Q^{t+n} at iteration levels t and t+n, respectively. Here, n is the number of regular iterations performed by the original non-accelerated algorithm. Then, ΔQ between the two iteration levels is defined as $Q^{t+n} = Q^t + \Delta Q$ if no acceleration algorithm is used. Using the first two terms of a Taylor series expansion in the artificial (iterating) time direction, the residual for each of the equations in the system after n iterations is

$$R_r^{t+n} = R_r^t + \sum_m^M \sum_s^S \frac{\partial R_r^t}{\partial Q_{m,s}} \Delta Q_{m,s} \qquad (98)$$

Notice that the total number of equations in the system is the same as the total number of unknown components of Q, that is, rmax = M. If we introduce convergence rate acceleration coefficients α_1, α_2, ... , α_M multiplying corrections respectively, then each component, $Q_{m,s}^{(t+n)+1}$, of the future solution vector at the grid point, s, can be extrapolated as

$$Q_{m,s}^{(t+n)+1} = Q_{m,s}^t + \alpha_m \Delta Q_{m,s} \qquad (99)$$

This can be applied at every grid point in the domain, but it results in a huge system of imax x jmax x kmax x M unknown acceleration coefficients, α. If each of the α's is assumed to have the same value over the entire domain, D, the number of unknown a's is reduced to M. This pragmatic approach is called the global SBMR method [311], [312]. It requires solution of a much smaller M x M matrix. The future residual at the iteration level (t+n)+1 can, therefore, be approximated by

$$R_r^{(t+n)+1} = R_r^t + \sum_m^M \left[\alpha_m \cdot \sum_s \frac{{}^s \partial R_r^t}{\partial Q_{m,s}} \Delta Q_{m,s} \right] \tag{100}$$

Subtracting (94) from (96) yields

$$R_r^{(t+n)+1} = R_r^t + \sum_m^M (\alpha - 1) \cdot a_{rm} \tag{101}$$

where

$$a_{rm} = \sum_s \frac{{}^s \partial R_r^t}{\partial Q_{m,s}} \Delta Q_{m,s} \tag{102}$$

The optimum α's are determined such that the sum of the L-2 norm of the future residuals over the entire domain, D, will be minimized.

$$\sum_D \sum_r^{rmax} \frac{\partial (R_r^{(t+n)+1})^2}{\partial \alpha_m} = 2 \sum_D \sum_r^{rmax} \frac{\partial R_r^{(t+n)+1}}{\partial \alpha_m} R_r^{(t+n)+1} = 0 \tag{103}$$

for m = 1,..,M. With the help of (101) and (102), the system (103) becomes

$$\sum_D \sum_r^{rmax} \left\{ R_r^{t+n} + \sum_m^M a_{rm} \cdot (\alpha - 1) a_{r1} \right\} = 0$$

$$\sum_D \sum_r^{rmax} \left\{ R_r^{t+n} + \sum_m^M a_{rm} \cdot (\alpha - 1) a_{r2} \right\} = 0 \tag{104}$$

$$\cdots$$

$$\sum_D \sum_r^{rmax} \left\{ R_r^{t+n} + \sum_m^M a_{rm} \cdot (\alpha - 1) a_{rM} \right\} = 0$$

In equation (101), the R's and a's are known from the preceding iteration levels. Since each α_m is assumed to have the same value over the entire computational domain, equation (104) gives a tractable system of M simultaneous algebraic equations for M optimum $\alpha_1, \alpha_2, ...,$ α_M.

$$\left[\sum_D\left(\sum_r^{rmax} a_{r1}a_{r1}\right)\right]\cdot(\alpha_1-1)+...+\left[\sum_D\left(\sum_r^{rmax} a_{rM}a_{r1}\right)\right]\cdot(\alpha_M-1) = -\sum_D\left(\sum_r^{rmax} R_r^{t+n}a_{r1}\right)$$

$$\left[\sum_D\left(\sum_r^{rmax} a_{r1}a_{r2}\right)\right]\cdot(\alpha_1-1)+...+\left[\sum_D\left(\sum_r^{rmax} a_{rM}a_{r2}\right)\right]\cdot(\alpha_M-1) = -\sum_D\left(\sum_r^{rmax} R_r^{t+n}a_{r2}\right) \quad (105)$$

$$...$$

$$\left[\sum_D\left(\sum_r^{rmax} a_{r1}a_{rM}\right)\right]\cdot(\alpha_1-1)+...+\left[\sum_D\left(\sum_r^{rmax} a_{rM}a_{rM}\right)\right]\cdot(\alpha_M-1) = -\sum_D\left(\sum_r^{rmax} R_r^{t+n}a_{rM}\right)$$

For the general case of a system composed of M partial differential equations with M unknowns, the system (105) will become a full M x M symmetric matrix for M unknown optimum α's.

It is plausible that for non-uniform computational grids and rapidly varying dependent variables, optimum α's should not necessarily be the same over the whole computational domain. A modification of the SBMR method called Line Sensitivity-Based Minimal Residual (LSBMR) method was developed [311], [312] to allow α's to have different values from one grid line to another. The resulting system has, for example, jmax x M unknown α's, which is quite tractable, although it is more complex to implement than the SBMR method.

The performance of the SBMR and the LSBMR methods [311], [312] depends on how frequently these methods are applied during the basic iteration process and on the number of iterations performed with the basic iterative algorithm that are involved in the evaluation of the change of the solution vector. In the case of two-dimensional incompressible viscous flows without severe pressure gradient, the SBMR and LSBMR methods significantly accelerate the convergence of iterative procedure on clustered grids with the LSBMR method becoming more efficient as grids are becoming highly clustered. The SBMR and the LSBMR methods tested for a two-dimensional incompressible laminar flow maintain the fast convergence for highly non-orthogonal grids and for flows with closed and open flow separation. The SBMR method is capable of accelerating the convergence of inviscid, low Mach number, compressible flows where the system is very stiff. An Alternating Plane Sensitivity-Based Minimal Residual (APSBMR) method [312], a three-dimensional analogy of the LSBMR method, has been shown to successfully reduce the computational effort by 50% when solving a three-dimensional, laminar flow through a straight duct without flow separation.

The general formulation of these acceleration methods is applicable to any iteration scheme (explicit or implicit) as the basic iteration algorithm. Hence, it should be possible to apply these convergence acceleration methods in conjunction with the other iteration algorithms and with other acceleration methods (preconditioning [307], multigridding [310], etc.) to explore the possibilities for a cumulative convergence acceleration effect.

In the case of optimization methods based on sensitivity analysis, the main difficulty is to make the evaluation of the derivatives less difficult and more reliable. A worthwhile effort is to research further on use of the automatic differentiation (ADIFOR) based on a chain-rule for evaluating the derivatives of the functions defined by flow analysis code with respect to its input variables [313]-[317]. This approach still needs to be made more computationally efficient. One obvious possibility is to utilize parallel computer architecture in the optimization process not only with the gradient search algorithms [318], [319], but especially with genetic evolution algorithms [320] where massively parallel approach should offer impressive reductions in computing time. The use of parallel computing will become unavoidable as the scope of optimization becomes multiobjective [321]-[325].

An equally worthwhile effort is to further research the "one shot" method that carefully combines the adjoint operator approach and the multigrid method [326]. This algorithm optimizes the control variables on coarse grids, thus eliminating costly repetitive flow analysis on fine grids during each optimization cycle. The entire shape optimization should be ideally accomplished in one application of the full multigrid flow solver [326].

Finally, it should be pointed out that although the exploratory efforts in applications of neural networks [327], [328] and virtual reality are still computationally inefficient, their time is coming inevitably as we attempt to optimize the complete real aircraft systems. Recent pioneering efforts [329] in coupling wind tunnel experimental measurements and a multiobjective hybrid genetic optimization algorithm indicate that the decades old dream of effectively and harmoniously utilizing both resources is soon to become a reality.

15.4 Parallelization and Networking Issues

The treatment of complex geometries has led us to adopt a multi-block grid made of several structures with overlapping or patched domains. The method of distributing domains over processors is important in that it leads to typical load balancing problems and synchronization of waiting time. Issues that need to be addressed are: flexibility in load balancing, reading and generating block data, reading and generating interface data, and updating block and interface data during communication. The advantage in using the multi-block approach is that one can have more than one block solver (for example, Euler, Navier-Stokes, etc.) in different blocks depending on the complexity of the flow field in a given block, thereby improving the overall efficiency of the algorithm. The most important parallelization issue for CFD applications is the way in which the computational domain is partitioned among a cluster of processors. Even a highly efficient parallel algorithm can give poor results for a poorly implemented domain decomposition. The domain decomposition technique has been successfully implemented on both SIMD and distributed memory MIMD computers. Algorithms like the "Masked Multi-block Algorithm" allow for dynamically partitioning the domain depending on the distribution of load among processors. This eliminates the possibility of distributing each domain on a processor since in most cases domains will have irregular sizes. Other algorithms distribute separated planes of a 3-D computational domain between processors to synchronize time waiting. The measure of the efficiency of a parallel algorithm is given by the ratio of the computation time to the communication time for a particular application. High performance message passing can be achieved by "overlapping communication", performing assembly-coded gather-scatter operations. A machine like the CM-2 is a SIMD type machine where most of the parallelization is carried out by the compiler which is responsible for data layout. The Cray's parallel processing capabilities can be exploited by the use of "auto-tasking" wherein the user indicates points of potential parallelism in the implementation by the use of directives which instruct a pre-processor to reconfigure the source program in such a way so as to enable maximum speedup to be obtained.

An alternative to parallel hardware architecture is the poor man's machine or PVM (Parallel Virtual Machine) that can simulate a parallel machine across a host of serial machines. The programming model supported by PVM is distributed memory multi-processing with low level message passing. A PVM application essentially uses routines in the PVM to do message passing, process control and automatic data conversion. A special process runs on each node (each machine) of the virtual machine and provides communication support and process control. However since message passing is carried out on the Ethernet it is considerably slower than the Intel Interprocessor network. There is also an overhead on account differences between speeds of different machines that create load balancing problems.

15.5 Suggested General Structure of the MIDO Algorithms

Any MIDO algorithm stresses high transportability as well as extensive graphical user interaction and support. The product is a full featured, self-configuring MIDO package capable of utilizing a variable number of diverse CPU's and input/output devices that is valuable to research and industrial practical use. Massively parallel machines are the latest significant step in this technology. This expansion from a single CPU has greatly increased computational power; however, it has also increased the problem of transportability of the software. The rationale behind the advent of FORTRAN 90 was that users needed a standard method for the utilization of a variable number of CPU's. The CPU job tasking, while referred to by the programmer, is essentially handled by the FORTRAN 90 compiler. While FORTRAN 90 delivers a standard to the user of parallel machines, it in no way increases the user's interface with the code itself. The researcher or design engineer who simply provides a source code to the FORTRAN 90 compiler has no standard method for producing runtime graphics, real-time flow-field visualization, or other graphical user interactions. The only option for the user is to make calls to an external graphics package that is loaded on the particular system in use. By doing this, the generality of the code is lost. This problem is also true of other input/output devices.

This problem can be solved by giving the user real-time graphical visualization as an integral part of the MIDO software package that will allow researchers and design engineers to watch as they analyze and optimize their designs. Furthermore, criteria and boundary conditions are dynamic, allowing users to alter or update them as execution progresses. The problem of transportability is overcome because the package does not require a compiler since such MIDO package is pre-compiled and pre-linked. It is a stand-alone executable that detects and utilizes the host system. It is a self-contained, auto-configuring package that is to aerodynamic shape design and thermal design what I-DEAS and GENESIS are to structural design. An executable such as this MIDO package, can also be developed in lower level languages such as C/C++ or Assembler rather than less efficient high level languages such as FORTRAN. As well as being more efficient, low level languages operate in a regime that is close to the native language of the machine, allowing direct manipulation of the host system hardware. This direct access programming allows the final product to be highly modular and adaptable. These advantages could be utilized in the MIDO package with the end result that the user can choose the desired aspects of the package, and the algorithm will conform itself.

Because of the complex algorithms and high degree of programming skills required for this package, the development process can be divided in two stages.

During the first stage of such a MIDO package development, it can be written in FORTRAN 77 so that its legitimacy can be checked. Numerical solutions produced by this package can be checked against available analytical and experimental data to ensure that true physical phenomena are captured. Then, the MIDO package can be rewritten into low level languages as modules to a central-switching logic that is designed for a single CPU system. A full-featured interactive graphical environment can be developed for this system.

The second stage of the suggested MIDO package development can be centered on the expansion of the package to include support for a variable CPU system. Package development can culminate in its high modularity, full user interface environment (for a PC with MS-Windows-LINUX and workstations with UNIX and X-Windows), ability to change flow solvers in the environment, change boundary conditions, switch between analysis mode, inverse design mode, and optimization (based on user defined criteria) mode in the environment while developing a desired design.

Although most aerodynamic, heat transfer and elasticity analysis codes are written in FORTRAN while genetic algorithms and graphical interfaces are more efficient when coded in C++ language, the use the PVM (Parallel Virtual Machine) or the new generation MPL (Message Passing Libraries) can make their simultaneous use possible. With the PVM or MPL, the hybrid GA optimizer can be compiled from a C++ source, while an analysis code can be compiled from a FORTRAN source. The message-passing between the two running codes is not extensive, yet it is independent of the language of the source code.

The efforts should be focused in developing a commercial quality MIDO package for research and direct industrial use. Algorithms should be modified to run in FORTRAN 90 on parallel CM-2, CM-5, IBM SP-2, CRAY-90, etc. or clusters of arbitrary PC's, workstations and mainframes connected through a router thus stressing transportability to an arbitrary number of CPU's.

15.6 Conclusions

The Multidisciplinary Inverse Design and Optimization (MIDO) approach to a complete aircraft system design is already a reality in some of the leading companies. It has been taking two distinct paths where either a semi-sequential disciplinary optimization is performed or a simultaneous analysis and optimization of the entire system is performed. Both approaches have their own difficulties with computing time requirements and numerical stability. These issues will progressively be resolved by a better use of massively and distributed parallel computation and by the development of faster iterative algorithms for analysis and constrained optimization.

15.7 References

[289] **Sobieski, J. S.**
Multidisciplinary Optimization for Engineering Systems: Achievements and Potential, NASA TM 101566, March 1989.

[290] **Voigt, R. G.**
Requirements for Multidisciplinary Design of Aerospace Vehicles on High Performance Computers, ICASE Report No. 89-70, Sept. 1989.

[291] **Dovi, A. R., Wrenn, G. A.**
Aircraft Design for Mission Performance Using Nonlinear Multiobjective Optimization Methods, J. of Aircraft, Vol. 27, No. 12, 1990, pp. 1043-1049.

[292] **Savu, G., Trifu, O.**
On A Global Aerodynamic Optimization of a Civil Transport Aircraft", Proc. of 3rd Int. Conf. on Inverse Design Concepts and Optimiz. in Eng. Sci. (ICIDES-III), editor: G.S. Dulikravich, Washington, D.C., October 23-25, 1991.

[293] **Fornasier, L.**
Numerical Optimization in Germany: A Non-Exhaustive Survey on Current Developments with Emphasis on Aeronautics, Proc. of 3rd Int. Conf. on Inverse Design Concepts and Optimiz. in Eng. Sci. (ICIDES-III), editor: Dulikravich, G. S., Washington, D. C., October 23-25, 1991.

[294] **Sobieszczanski-Sobieski, J., Haftka, R. T.**
Multidisciplinary Aerospace Design Optimization: Survey of Recent Developments, AIAA paper 96-0711, Reno, NV, January 1995.

[295] **Appa, K., Argyris, J.**
Nonlinear Multidisciplinary Design Optimization Using System Identification and Optimal Control Theory, AIAa-95-1481, New Orleans, LA, April 1995.

[296] **Arian, E.**
Analysis of the Hessian for Aeroelastic Optimization, ICASE Report No. 95-84, Dec. 1995.

[297] **Kennon, S. R., Dulikravich, G. S.**
A Posteriori Optimization of Computational Grids, AIAA Paper 85-0483, Reno,NV, 1985; also AIAA Journal, Vol. 24, No. 7, July 1986, pp. 1069-1073.

[298] **Carcaillet, R., Kennon, S. R., Dulikravich, G. S.**
Optimization of Three-Dimensional Computational Grids, AIAA Paper 85-4087, Colorado Springs, CO, October 1985; also Journal of Aircraft, Vol. 23, No. 5, May 1986, pp. 415-421.

[299] **Kennon, S. R., Dulikravich, G. S.**
Composite Computational Grid Generation Using Optimization, Proceedings of the First International Conf. on Numerical Grid Generation Comp. Fluid Dynamics, editor J. Haeuser, Landshut, W. Germany, July 14-17, 1986.

[300] **Carcaillet, R., Dulikravich, G. S., Kennon, S. R.**
Generation of Solution Adaptive Computational Grids Using Optimization, Computer Meth. in Appl. Mech. and Eng., Vol. 57, Sept. 1986, pp. 279-295.

[301] **Jameson, A., Schmidt, W., Turkel, E.**
Numerical Solutions of the Euler Equations by Finite Volume Methods Using Runge-Kutta Time-Stepping Schemes, AIAA Paper 81-1259, June 1981.

[302] **Martinelli, L., Jameson, A., Grasso, F.**
A Multigrid Method for the Navier-Stokes Equations, AIAA Paper 86-0208, January 1986.

[303] **Chima, R. V. and Turkel, E., Schaffer, S.**
Comparison of Three Explicit Multigrid Methods for the Euler and Navier-Stokes Equations, AIAA Paper 87-0602, January 1987.

[304] **Saad, Y., Schultz, M.**
Conjugate Gradient-Like Algorithms for Solving Non-symmetric Linear Systems, Mathematics of Computation, Vol. 44, No. 170, 1985, pp. 417-424.

[305] **Hafez, M., Parlette, E., Salas, M. D.**
Convergence Acceleration of Iterative Solutions of Euler Equations for Transonic Flow Computations, AIAA Paper 85-1641, July 1985.

[306] **Huang, C. Y., Dulikravich, G. S.**
Fast Iterative Algorithms Based on Optimized Explicit Time-Stepping, Computer Methods in Applied Mechanics and Engineering, Vol. 63, August 1987, pp. 15-36.

[307] **Turkel, E.**
Review of Preconditioning Methods for Fluid Dynamics, ICASE Report No. 92-47, NASA Langley Research Center, Hampton, VA, Sept. 1992.

[308] **Lee, S., Dulikravich, G. S.**
Distributed Minimal Residual (DMR) Method for Acceleration of Iterative Algorithms, Computer Methods in Applied Mechanics and Engineering, Vol. 86, 1991, pp. 245-262.

[309] **Lee, S., Dulikravich, G. S.**
Accelerated Computation of Viscous Flow with Heat Transfer, Numerical Heat Transfer: Fundamentals, Part B, Vol. 19, June 1991, pp. 223-241.

[310] **Brandt, A.**
Multigrid Technique, Guide with Applications to Fluid Dynamics, GMD Studien 85, GMD-AIW, Postfach 1240, D-5205 St. Augustine, Germany, 1984.

[311] **Choi, K. Y., Dulikravich, G. S.**
Sensitivity-Based Methods for Convergence Acceleration of Iterative Algorithms, Computer Methods in Applied Mechanics and Engineering, Vol. 123, Nos. 1-4, June 1995, pp. 161-172.

[312] **Choi, K. Y., Dulikravich, G. S.**
Acceleration of Iterative Algorithms on Highly Clustered Grids, AIAA Journal, Vol. 34, No. 4, April 1996, pp. 691-699.

[313] **Griewank, A., Corliss, G. F. (editors)**
Automatic Differentiation of Algorithms: Theory, Implementation and Application, SIAM, Philadelphia, PA, 1991.

[314] **Bischof, C. H., Carle, A., Corliss, G., Griewank, A., Hovland, P.**
ADIFOR - Generating Derivative Code from Fortran Programs, Scient. Programming, Vol. 1, No. 1, 1992, pp. 1-29.

[315] **Green, L. L., Newman, P. A., Haigler, K. J.**
Sensitivity Derivatives for Advanced CFD Algorithm and Viscous Modelling Parameters via Automatic Differentiation, AIAA paper 93-3321-CP, Orlando, FL, July 6-9. 1993.

[316] **Hou, G. J.-W., Maroju, W., Taylor, A. C., Korivi, V., Newman, P. A.**
Transonic Turbulent Airfoil Design Optimization With Automatic Differentiation in Incremental Iterative Forms, AIAA CFD Conference Proceedings, San Diego, CA, June 1995, AIAA-95-1692-CP, pp. 512-526.

[317] **Hovland, P., Altus, S., Kroo, I., Bischof, C.**
Using Automatic Differentiation With the Quasi-Procedural Method for Multidisciplinary Design Optimization, AIAA paper 96-0090, Reno, NV, January 1996.

[318] **Cheung, S.**
Aerodynamic Design Parallel CFD and Optimization Routines, AIAA CFD Conference Proceedings, San Diego, CA, June 1995, AIAA-95-1748-CP, pp. 1180-1187.

[319] **Jameson, A., Alonso, J. J.**
Automatic Aerodynamic Optimization on Distributed Memory Architectures, AIAA paper 96-0409, Reno, NV, January 1996.

[320] **Poloni, C., Fearon, M., Ng, D.**
Parallelisation of Genetic Algorithm for Aerodynamic Design Optimisation, Proceedings of 2nd Conf. on Adaptive Computing in Engineering Design and Controls (ACEDC), Plymouth, UK, March 23-27, 1996.

[321] **Osyczka, A.**
Multicriterion Optimization in Engineering, Ellis Hoewood, Ltd., England, 1984.

[322] **Eschenauer, H., Koski, J., Osyczka, A.**
Multicriteria Design Optimization Procedures and Applications, Springer-Verlag, Berlin, 1990.

[323] **Azarm, S., Sobieszczanski-Sobieski, J.**
Reduction Method With System Analysis for Multiobjective Optimization-Based Design, ICASE Report No. 93-22, April 1993.

[324] **Poloni, C., Mosetti, G., Contessi, S.**
Multi-Objective Optimization by GAs: Application to System and Component Design, invited STS lecture at ECCOMAS 96 (European Community in Computational Methods in Applied Sciences), Paris, France, Sept. 9-13, 1996.

[325] **Poloni, C.**
Hybrid Genetic Algorithm for Multiobjective Aerodynamic Optimisation, in: Genetic Algorithms in Engineering and Computer Science, pp. 397-415, John Wiley & Sons, England, December 1995, ISBN 0-471-95859-X, Contributed book edited by: G. Winter, J. Periaux, M. Galan and P. Cuesta.

[326] **Ta'asan, S.**
Trends in Aerodynamics Design and Optimization: A Mathematical Viewpoint, AIAA CFD Conference Proceedings, San Diego, CA, June 1995, AIAA-95-1731-CP, pp. 961-970.

[327] **Prasanth, K., Markin, R.E., Whitaker, K. W.**
Design of Thrust Vectoring Exhaust Nozzles for Real-Time Applications Using Neural Networks, Proc. of 3rd Int. Conf. on Inverse Design Concepts and Optimization in Engineering Sciences (ICIDES-III), editor: G. S. Dulikravich, Washington, D. C., October 23-25, 1991.

[328] **Huang, S. Y., Miller, L. S., Steck, J. E.**
An Exploratory Application of Neural Networks to Airfoil Design, AIAA paper 94-0501, Reno, NV, January 1994.

[329] **Poloni, C., Mosetti, G.**
Private Conversation, April 1996.

CHAPTER 16

SUPERSONIC AIRCRAFT SHAPE OPTIMIZATION

A. Van der Velden

Synaps Inc., Atlanta, GA, USA

16.1 Abstract

This paper will discuss examples of supersonic aerodynamic shape optimization as developed for Daimler-Benz Aerospace Airbus by Synaps, Inc.

First, we will introduce a general approach to aerodynamic shape design based on minimization of energy consumption during aircraft life while considering realistic constraints on lift, pitching and rolling moments and geometric dimensions. The analysis is performed using a potential code with real flow corrections and a decoupled boundary layer calculation. Finally, this method is applied to the design of the European Supersonic Civil Transport and the Oblique Flying Wing.

16.2 Introduction

Engineers have long sought to improve wing design methods. Initially, simple geometric shape functions were used to characterize airfoil shapes; the NACA airfoil classification system using this method is described in reference [330]. Unfortunately these NACA airfoils had high drag at transonic speeds, and therefore more refined shapes with reduced (or no) transonic shocks were required. More recently, successful transonic wings have been designed by defining a (nearly) shock-free transonic pressure distribution and using an inverse solver to find the corresponding airfoil geometry. Though such methods can be applied in high-transonic flow, in supersonic flow these methods loose their meaning because the optimal pressure distribution cannot be shock-less.

During the days of Concorde and the SST, methods not based on pressure distributions were introduced to solve supersonic aircraft design problems. These methods are still in use to-day. A good overview paper on these methods was written by Baals, Robins and Harris [331]. The method first area-rules the fuselage wing combination. The wing lift distribution is opti-mized with Langrange's method of undetermined multipliers [349][334] for minimum drag. This optimum lift distribution can then be inverted into the geometry using linear theory. But often, the wing shapes derived from such inverse methods exhibit camber kinks due to ill conditioned aerodynamic matrices that require post processing. Variations in the wing planform can also not be considered since it would require a recomputation of the aerodynamic matrices.

A solution that can be used for both transonic and supersonic flow was proposed by VanderPlaats and Hicks [341]. They applied numerical optimization techniques combined with a nonlinear flow analysis methods such as Euler to modify a wing shape such that an arbitrary objective function representing the design goals could be minimized. Although their work was done twenty years ago, computer speeds have not increased sufficiently to make it practical with the shape functions they proposed. Recently Hicks, Reuther and Jameson have applied this tech-nique on supersonic wing body combinations using the Euler equations.

Since the underlying potential flow theory has appeared to work very well for prelimi-nary supersonic aircraft design, most manufacturers therefore see no need to change to compu-tationally more expensive and less tested methods at this time. Another advantage of potential flow cited by some engineers involved in the ESCT project, was the inherent separation of wave drag, induced drag and friction drag. This separation of drag components helped to analyze prob-lem spots and the low computational effort reduced the critical design turn around time. Though the author agrees with this view, the design method as practiced today is cumbersome and in-flexible as to the use of constraints and modeling of new types of configurations.

The current industrial supersonic wing design methods typically do not include leading edge suction criteria or vortex development. Both phenomena influence the forces and moments significantly, especially at subsonic speeds.

This paper combines the direct numerical optimization of a completely flexible wing-body-nacelle configuration with an industrial potential flow code with corrections.

16.3 General Methodology

Multipoint aerodynamic optimization can generally be described in the form of reference [353]:

1. Select objective function.

2. Select variables and constraints.

3. Select optimizer(s).

4. Optimize and analyze the results.

16.3.1 Drag as an Objective Function

Consider the objective function O described by the total energy loss due to aircraft drag during the operation of the aircraft over n missions:

$$O = \sum_{i=1}^{n} p_i \int_{x=0,i}^{r_i} c_D q S dx \qquad (106)$$

In this expression, p_i is the probability of mission i, and r_i is the block range for that mission. This expression can be simplified by weighing the drag over a number of representative design points:

$$O = \sum_{i=1}^{m} w_i c_D \qquad (107)$$

In our experience, five design points are enough to describe a multipoint design adequately.

16.3.2 Variables

The variables of our problem describe the aerodynamic shape of the aircraft. Over the last two decades, many engineers have tried to describe the shape of a wing accurately with only a few parameters while maintaining flexibility. There are three general classes of aerodynamic shape functions:

- *Linear combination* of existing wing sections. This geometric approach to the airfoil design problem is a derivative of the original NACA method [330]. Although this method is computationally efficient, at DA this method has often given only trivial solutions (*e.g.* the old airfoil) due to the low flexibility inherent to this geometrical representation. In addition, there is little physical foundation for the implicit assumption that the drag (objective) of a linear combination of airfoil geometries is linearly or at all related to the drag of the individual airfoils.

- *Analytical* shape functions are linearly superimposed to define a wing geometry. Examples of these include Hicks-Henne functions, Wagner functions and the patched polynomials discussed in reference [344], as well as the Legendre polynomials used by Reneaux [346] and the splines used by Cosentino [336]. Unfortunately, the number of parameters required for these shape functions are usually too high to allow multipoint three-dimensional design. Every time, a minimum of 30–40 points are required to define each airfoil. Even if only five airfoils are sufficient for defining a wing, about 200 parameters will be necessary. This number of variables is simply too great for practical optimization at present. Cosentino optimizes a 3D wing by allowing variations over only a small portion of the wing. Lee has proposed patched polynomials as a way of flexiblely modeling an airfoil section with only sixteen parameters. Although this approach does produce a flexible geometry, the performance of the sections designed with this method cannot compete with those produced by experienced engineers using inverse methods.

- *Special aero-functions* were proposed by Reneaux [346] for designing airfoils with a minimum number of parameters. This approach automates the steps that an experienced designer follows when using an inverse method. 'Good' pressure distribution types are defined, along with the shape or special aero-functions that produce these pressure distributions. Unfortunately, for a given pressure distribution type, these aero-functions vary with Mach number, and this method is therefore impractical for multipoint designs. This method has all the disadvantages of inverse methods discussed in the Introduction.

At DA [352] the author has introduced another type of function, a highly nonlinear special aero-function, which can optimize the shape of a 3D wing with adequate flexibility using current computer technology. It is now being used for the industrial design of transonic and supersonic wings. Its application will be shown in the next section. These shape functions produce the aircraft geometry in any resolution.

16.3.3 Constraints

The type and number of constraints are highly problem-dependent. Typical constraints for commercial wing designs are:

- *Lift and angle of attack.* Either the lift C_L or angle of attack α are constrained or defined as input values. For the horizontal tail plane, both can be simultaneously constrained.

- *Minimum and maximum lift.* Maximum lift at low speed ($\approx$ 50m/s) $C_{L,max}$ for a given Reynolds number influences takeoff and landing performance and is therefore an important parameter. For horizontal tails, the minimum lift at low speed $C_{L,min}$ is an important constraint. For slender supersonic aircraft the maximum lift is usually defined by a maximum angle of attack of 20°. according to the 'Concorde' SST regulations.

- *Buffet onset* at 1.3g relative to 1g cruise. Although a buffet onset is difficult to predict exactly, a conservative estimate can be made using the trailing edge separation criterion [340] based on trailing edge pressure coefficients. Buffet of this type is usually a problem for transonic wing designs.

- *Pitching moment C_m.* This is usually not particularly important for subsonic transports but is a major issue for supersonic transports. The hinge moment $C_{m,h}$ is an important constraint in the design of control surfaces.

- The aircraft *thrust must be at least equal to the drag* also at off-design points, of which the most critical are usually at the high lift and high Mach number corner of the envelope. The allowed drag creep as a function of Mach number or lift is also often constrained.

- *Geometry.* The airfoil thickness and thickness distribution are usually subject to structural and geometric constraints such as spar depths.

- *Engine installation.* The wing design should take into account the disturbances due to engine installation. This is done implicitly by including the engine in- and outflow with the nacelle in the drag balance. When a surface (such as a wing) is placed near a nacelle it will then assume a shape such that the drag is minimized (e.g. interference minimized)

16.4 Analysis

A linear panel code with real flow corrections was used to analyze drag for 3D supersonic wing body configurations. This analysis code was developed by the author from Woodward's wing-body code [355], and solves the linear Prandtl-Glauert equation for a thin panel geometry. The power plants are modeled as stream tubes that displace ambient fluid and therefore cause wave drag. Friction is calculated stripwise over the curved geometry assuming a fully turbulent boundary layer. No effort was made to couple the boundary layer equation to the potential flow pressure distribution. Potential drag is calculated with surface integration and a calculated value of the leading edge thrust. The correct leading edge thrust is found by combining the far-field drag as calculated by the Lomax supersonic area rule [345] and Trefftz-plane integration. The leading edge suction is then corrected for real flow effects using the method of Carlson [335]. The Torenbeek quasi-empiric method [348] is used to model the effect of flap deflection. An overview of this method is given in Figure 91. A more detailed description can be found in the author's thesis [350].

The forces and moments calculated with this method compare well with values measured from the NASA Ames Oblique Wing development program, a generic SCT, a generic fighter and the Munroe NASA arrow wing tests. Figure 92 shows the calculated and measured forces and moments for a generic SCT with 12 degrees leading and 3 degrees trailing edge flap deflection at Mach 1.05—a difficult case for panel calculations. Our method, which was applied to a panel half-model with a resolution of no more than 120 panels, took one second per flow point on our workstations. The Euler method EUFLEX [343] with a decoupled boundary layer calculation took about an hour on a Cray.

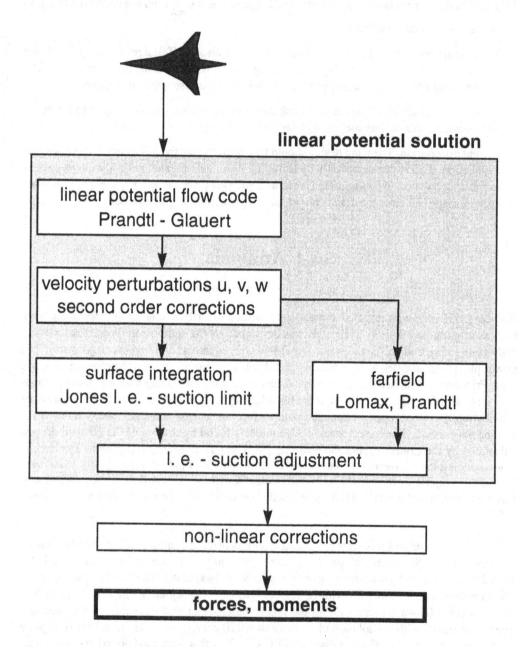

Figure 91   **A panel method with flow corrections to evaluate forces.**

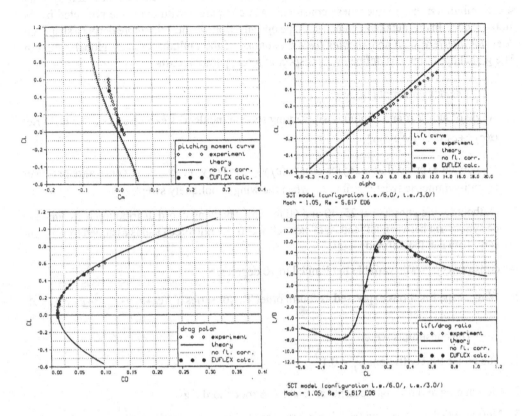

Figure 92   Forces and moments of a generic SCT.

16.5 Examples

16.5.1 Oblique Flying Wing

The design presented here is part of a MIDAS design cycle described previously in paper [354]. The global design was optimized for a Mach number of 1.6 and therefore had a fixed wing planform, sweep and thickness distribution. In the global model a preliminary wing planform was determined, and the assumption was made that there existed a thickness and camber distribution which would produce a lift-to-drag ratio of 11 at Mach 1.6 with a cruise lift coefficient of 0.10. This globally optimized aircraft had a span of 120m and a 14.2 m wide center section chord that was 2.7 m thick. It was assumed that near elliptic off design load distributions could be obtained by unsweeping the wing at lower Mach numbers and deflecting the trailing edge suitably. It is therefore acceptable to just optimize the aircraft with respect to cruise drag while keeping the span, and the center section (passenger cabin) dimensions the same.

The objective of the aerodynamic shape optimization is to find the best aerodynamic shape (minimum drag) under these constraints and compare it with the value projected by the global model. In this detailed geometric shape optimization we must therefore derive a shape that *does not interfere* with and *enables the assumptions* of the multidisciplinary global optimum. The global optimum was defined with respect to total operating costs.

The design variables were:

1. Angle of attack α and continuously varying section incidences expressed as a Taylor series $\alpha(c_1 y + c_2 y^2 + c_3 y^3 + c_4 y^4)$

2. Wing bend expressed as a Taylor series $z(c_1 y + c_2 y^2 + c_3 y^3 + c_4 y^4)$. The z curvature produces an anti-symmetric twist distribution when the wing is obliquely swept.

3. Outboard wing chord(y) at 25, 40 and 60 m (tip). The wing chord distribution is assumed to be symmetric between left and right side (all y cuts are mirrored) and is assumed to vary linearly between the defined y cuts.

4. cruise sweep Λ was variable but limited to 70 degrees.

5. Wing shape functions expressed as fourth order Taylor series $Y_i (c_0 + c_1 y + c_2 y^2 + c_3 y^3 + c_4 y^4)$

The constraints were:

1. The cruise lift coefficient $C_L < 0.1$

2. No pitching and rolling moments around 32 % root chord c.g.

3. Normal Mach numbers on the wing greater than 0.4 to prevent trailing edge separation. Normal Mach numbers not greater than 1.1 to prevent strong nonlinear effects that are not accounted for in the linear potential code.

4. Aisles higher than 2.2 m, cabin doors higher than 1.95 m, cargo hold higher than 1.7 m and underfloor crash zone higher than 50 cm. An extra 10 cm is reserved on each side of the geometry for structural reinforcements. This corresponds to current Airbus standards. Including structural dimensions this resulted in a maximum external thickness of 2.70 m.

Figure 93 shows the optimized design. Remarkable is the wing bend that the optimizer applied to create an elliptic lift distribution. Wing bend has the same effect as twist, but as the wing unsweeps, bend does not produce an asymmetric loading. In comparison to earlier designs by the author based on inverse methods [350] the current designs are much more symmetric and have flatter pressure distributions. The current design also achieves near the minimum theoretical drag at supersonic flow. It achieved the projected lift-to-drag ratio of 11 very accurately and therefore validated the overall approach. The kink in the 3rd section from the left is not caused by the shape functions, but by the kink in the planform. The sections and pressure distributions shown in Figure 93 are streamwise slices of the oblique flying wing.

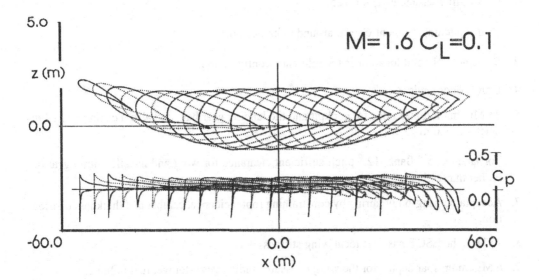

Figure 93   Optimized Oblique Flying Wing

16.5.2 Optimized European Supersonic Transport

The basis for the comparison was the already (fairly) optimal 1995 European Supersonic Transport Design. The goal of the optimization was to improve the cruise lift-to-drag ratio of the design by at least 5 % without penalizing the structural weight significantly and keeping the basic layout in terms of minimum dimensions. The engine nacelle was configured for cruise and not modified. The span was kept constant. The canard was considered to be not lifting at supersonic flight and was taken into account by a small drag increment.

The design variables were:

1. Angle of attack α, and twist angles expressed as a Taylor polynominal $\alpha_t(c_1y+c_2y^2+c_3y^3+c_4y^4)$. The root incidence was kept constant.

2. Wing location (x) at 33 %, 52 %, 86 % of the semi-span and the tip.

3. Wing chord(y) at 33 %, 52 %, 86 % of the semi-span and the tip. The root chord was kept constant.

4. Wing shape functions Y_i $(c_0+c_1y+c_2y^2+c_3y^3+c_4y^4)$

5. 19 Fuselage cross-sections as a function of x.

The following constraints were used:

1. Cruise lift coefficient $C_L < 0.125$,

2. Cruise pitching moment $C_m = 0$ around reference point.

3. Same neutral point location as the reference configuration.

4. Cruise floor angle $< 4.5^\circ$

5. 23 Minimum dimensions for cockpit and the landing gear bay, and also a cabin external diameter of 4.0 m ,

6. For landing: 5° Bank, 12° pitch sufficient clearance for wing and nacelle with a gear not higher than 5.5 m. ,

7. Acceptable wing-body fairing by constraining root incidence, camber and thickness distribution,

8. At least the ESCT-baseline local wing stiffness

9. 6 Minimum spar depths for the wing to assure landing gear storage, fuel volume.

The configuration was optimized for two operating points at Mach 2.0. Since the wing span and area were not affected it was not expected that the subsonic efficiency of the configuration would be changed after optimal flap scheduling was implemented.

The time required to reach a fully converged design was about three hours on a HP935/100Mhz. The original design goal was almost met, but with significant changes to the geometry. A free optimization would have achieved a greater reduction.

Figure 94 shows the optimized ESCT-6 as a coarse panel geometry. The optimizer swept the outboard panels more, but to keep the same stiffness as the reference ESCT, the thickness had to be increased. The optimizer also runs against the wing tip clearance constraint using the maximum allowable gear height. A solution was found by using forward sweep near the tip. This allowed the wing to be banked and pitched without striking the runway. As the eight wing section cuts show, the wing shape gradually changes from a rounded leading edge at the root to a sharp supersonic leading edge near the tip. The fuselage is significantly widened in the front. Although these changes were quite radical, the goal of 5% drag reduction was barely achieved, and it remains to be seen whether this new geometry has any other problems associated with it. In the optimization it turned out that the outboard wing sweep was the principle driving force in reducing the wing drag.

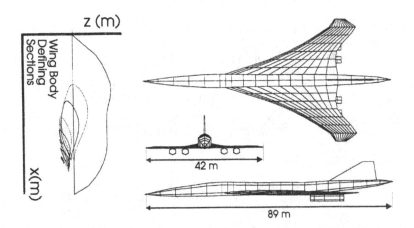

Figure 94   Optimized ESCT-6

16.5.3 Morphing a Supersonic into a Transonic Aircraft using Optimization

To illustrate the working of the optimizer we decided to change the Mach number from M 2.0 to 0.8 and make the wing span a design variable. In addition the length of the nacelles was halved to simulate a current high bypass ratio turbofan. We took the supersonic aircraft of the previous section and analyzed it at the subsonic speed of Mach 0.8. Though its lift-to-drag ratio improved to 12 because of the absence of wave drag, this was clearly not an optimal configuration at this flight condition. Figure 95 shows what happened. Immediately after the start of the optimization, the span is increased to the maximum allowable span of 60 m to reduce induced drag. Next the wing area is reduced to reduce friction drag. After about 12 hours, the configuration had converged to the last shape shown which had a lift-to-drag ratio of 22, slightly better than current transonic aircraft. The fact that the aircraft looks a bit like a VariEze probably stems from the fact that the landing gear is still stored in the wing and the aircraft has the same neutral point and pitching moment as the original ESCT. The outboard wing sweep looks fairly realistic. Local maximum wing normal Mach numbers are around 1.15, a very reasonable value for cruise flight, and perhaps a bit unexpected considering that we are running a panel code. Limitations to the leading edge suction and pressure levels as implemented in the potential flow code could be responsible for this phenomena. Area ruling is taken out and the minimum fuselage dimensions are selected. Notable is also the automatic repanneling of the wing-body interface with the wing root blended into the fuselage. If we look at the section design, we notice that wing looks like a transonic aircraft design: wash out twist, large leading edge radius, and rear loading.

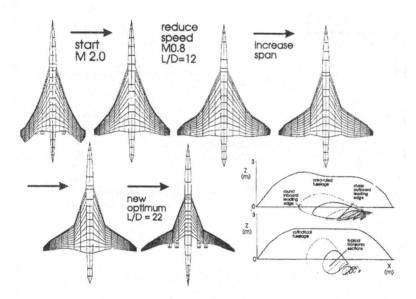

Figure 95   ESCT morphed into a VariEze type subsonic transport

Since shocks are not modeled here, and since the design constraints are not typical for a subsonic transport, one should not attach to much meaning to this morphed ESCT design. However, the results clearly show the flexibility of this method.

16.6 Conclusion

In this paper we present a general method for aerodynamic shape design, based on direct numerical optimization of aero-shape functions. The method was applied to the design of a 3D supersonic wing body configuration. This method is computationally efficient and competitive with designs obtained using current industrial methods, resulting in overall organizational time savings of approximately a factor of three and producing designs of comparable to better quality than those produced by inverse methods.

Acknowledgements

I would like to thank Ralph Carmichael of NASA Ames for his earlier help with the wingbody code, and my colleagues at Stanford University: Prof. Ilan Kroo, Peter Gage and Sean Wakayama, who helped me with my initial attempts to use optimization. Much of this work would not have been possible without the programming assistance of Silke Logemann. I would also like to thank Dr. M. Yoda for her editing of this paper. Finally, I would like to thank Dr. Mertens of Daimler Benz Aerospace Airbus for his organizational and personal support.

16.7 References

[330] **Abbott, I., von Doenhoff, A.**
Theory of Wing Sections, Dover Publications, New York 1949.

[331] **Baals, D. D., Robins, A., Harris, R.V.**
Aerodynamic Design Integration of Supersonic Aircraft, AIAA 68-1018, 1968

[332] **Bore, C.L.**
Propulsion Streamtubes in Supersonic Flow and Supercritical Intake Cowl, *Aeronautical J.*, September 1993.

[333] **Braunschädel, A.**
Erprobung Numerischer Aerodynamischer Verfahren für Überschallverkehrsflugzeuge an Verschiedenen Konfigurationen, Diplomarbeit Technische Hochschule Aachen, 1994.

[334] **Carlson, H.**
A Numerical Method For the Design of Camber Surfaces of Supersonic Wings With Arbitrary Planforms, NASA TND 2341, 1964

[335] **Carlson, H. W., Mack, R. J., Barger, R. L.**
Estimation of Attainable Leading-edge Thrust for Wings at Subsonic and Supersonic Speeds, NASA TP–1500, 1979.

[336] **Cosentino, G. M., Holst, T. L.**
Numerical Optimization Design of Advanced Transonic Wing Configurations, *J. Aircraft* **23** (3), 1986.

[337] **Dargel, G., Thiede, P.**
Viscous Transonic Airfoil Flow Simulation by an Efficient Viscous-inviscid Interaction Method, AIAA paper 87–0412, Jan. 1987.

[338] **Eminton, E., Lord, W. T.**
Note on the Numerical Evaluation of the Wave Drag of Smooth Slender Bodies Using Optimum Area Distributions for Minimum Wave Drag, *J. Roy. Aero. Soc.*, Jan. 1956.

[339] **Eppler, R., Somers, D.**
A Computer Program for the Design and Analysis of Low-speed Airfoils, NASA TM 80210, 1980.

[340] **Greff, E.**
In-flight Measurement of Static Pressures and Boundary Layer State with Integrated Sensors, *J. Aircraft* **28** (5), May 1991.

[341] **Hicks, R., Van der Plaats, G., Murman, E. M., King, R.**
Airfoil Section Drag Reduction at Transonic Speeds by Numerical Optimization, SAE Paper 760477, 1976.

[342] **Holst, T.**
Viscous Transonic Airfoil Workshop Compendium of Results, AIAA 87–1460, 1987.

[343] **Krämer, E., Gottmann, T.**
Berechnung des Strömungsfeldes eines Überschallflugzeugs mit Vorder- und Hinterkan-
tenklappen im Transschall mit Hilfe des EUFLEX - Verfahrens, DASA–LME211–S–
PUB–531, Münich 1993.

[344] **Lee, K. D.**
Application of Computational Fluid Dynamics in Transonic Aerodynamic Design,
AIAA paper 93–3481–CP, 1993.

[345] **Lomax, H.**
The Wave Drag of Arbitrary Configurations in Linearized Flow as Determined by Areas
and Forces in Oblique Planes, Ames Aeronautical Laboratory, NACA RM A55A18,
1955.

[346] **Reneaux, J.**
Numerical Optimization Methods for Airfoil Design, Recherche Aerospatiale no. 1984–
5, 1984.

[347] **Sobieczky, H.**
Progress in Inverse Design and Optimization in Aerodynamics, AGARD CP 463
Paper 1, (1989)

[348] **Torenbeek, E.**
Synthesis of Subsonic Airplane Design, Delft University Press, 1982.

[349] **Tucker, W.A.**
A Method for the Design of Sweptback Wings Warped to Produce Specified Flight Characteristics at Supersonic Speeds, NACA R 1226, 1956

[350] **Van der Velden, A. J. M.**
Aerodynamic Design and Synthesis of the Oblique Flying Wing Supersonic Transport,
Ph.D. Thesis, Stanford University, Dept. Aero./Astro. SUDAAR 621, UMI Microfilm
#DA9234183, June 1992.

[351] **Van der Velden, A. J. M.**
Multi-disciplinary SCT Design Optimization, AIAA paper 93–3931, 1993.

[352] **Van der Velden, A. J. M.**
Multi-point Optimization of Airfoils, Deutsche Aerospace Airbus Bericht EF–1979,
Nov. 1993 (not published).

[353] **Van der Velden, A., J. M.**
Tools for Applied Engineering Optimization, AGARD R 803, Apr. 1994.

[354] **Van der Velden, A. J. M., Von Reith, D.**
Multi-Disciplinary SCT Design at Deutsche Aerospace Airbus , 7th European Aerospace Conference EAC '94. 25-27 October 1994 Toulouse.

[355] **Woodward, F. A.**
Analysis and Design of Supersonic Wing-body Combinations, Including Flow Properties in the Near-field, NASA CR–73106, 1967.

CHAPTER 17

MULTI-DISCIPLINARY SUPERSONIC TRANSPORT DESIGN

A. Van der Velden
Synaps Inc., Atlanta, GA, USA

17.1 Abstract

The challenge in the development of a very complex system like a supersonic transport is not only to achieve the required technology, but also to link a team of highly skilled experts. In this paper a successful industrial approach is described to integrate the individual departments with their specific knowledge into the design of a future supersonic commercial transport.

Different designs are analyzed with a modular synthesis model and compared on the basis of operating economy with specified performance and environmental impact. The analysis routines of the synthesis model are mainly configuration independent and represent fixed levels of structural, aerodynamic and propulsion technology. The specialist departments are responsible for the content of the routines, and later verify the design with more refined methods. At present more than two hundred variables describe the aircraft geometry, engine characteristics and mission. Thirty of those variables representing the aircraft and its flight-profile are optimized simultaneously as a function of Mach number, payload and range. Because the various designs are analyzed with the same routines and optimization procedures they can be easily compared. This aircraft pre-optimization results in a significant reduction of the number of follow-on detail-design cycles, especially for non-conventional designs.

Examples are given for the preliminary design of arrow-wing and oblique wing supersonic aircraft as compared to subsonic aircraft using the same technology. It is also shown how technology and environmental constraints influence the sized design.

17.2 List of Principal Symbols

BPR	engine bypass ratio
C_L	Lift coefficient
DEM	design empty mass
h	altitude
IOC	Indirect Operating Costs
M	Mach number
M_{to}	maximum takeoff weight
l	length
L / D	lift-to-drag ratio
OFW	oblique flying wing
OWB	oblique wing body
S	reference wing area
SCT	Supersonic Civil Transport
s.f.c.	specific fuel consumption (N/hr/N)
SLS	sea level static
SWB	symmetric wing body
t/c	thickness to chord ratio
TOC	total operating cost per seat km
$T_{t,4,max}$	maximum turbine entry temperature
V_{mc}	minimum control speed
w	width

Greek Letters:

$\varepsilon_{c,max}$	maximum engine pressure ratio
Λ	sweep angle
ΔO_3	ozone depletion
ΔP	sonic boom sea-level overpressure

17.3 Introduction

In the early days of aviation, the technology to design aircraft was relatively simple and the requirements on product safety minimal. As a consequence, aircraft could be designed by small groups of people. Such small groups can communicate directly and therefore work very efficiently. For instance: In 1936 it took Kurt Tank exactly one year to conceive and produce the Focke Wulf Condor, the first transatlantic airliner. However as the technology became more complex, aircraft designers had to specialize to cope with the increased flow of information. In addition, the growing market required improved safety and accurate performance guarantees. Such performance and safety guarantees could only be made by extensive analysis and testing of the aircraft design. Due to this increased work-load an aircraft is no longer designed by a single group, but by hundreds of specialists in many departments. This subdivision of work further increased productivity and enabled the development of the complicated but safe transport aircraft we have today.

Although the specialists can fit the aircraft with the best technology available in their field, it is unclear whether this will always lead to the best aircraft. The best aircraft can only be designed with a truly interdisciplinary effort. The number of people and independent locations increases the design cycle time and decreases the amount of interaction between the disciplinary groups. Progress is thereby limited to incremental improvements making it difficult to achieve the breakthroughs in aircraft design still common thirty years ago. This paper will present a solution to this problem that was based on the author's thesis at Stanford University [371].

17.4 Overview

We will first discuss a design strategy which finds the optimum aircraft design in the analysis parameter space for a given mission. This parameter space is determined by the various disciplinary groups. The individual disciplines supply robust and physically correct modular analysis methods and the enabling technologies. Non-linear optimization techniques are then used to find the best design using these modules in this parameter space. This strategy reduces the number of design cycles and allows us to evaluate more configurations.

This strategy is applied to the design of supersonic transports. The main assumptions of the analysis routines will be described as well as the objective function. Several near-term technology supersonic configurations will be compared as a function of their mission specifications.

17.5 MIDAS: A Design Process

How doe we best achieve flexibility and efficiency in our design process while effectively using the talent available to the organization? Figure 96 shows our solution called **Multi-Disciplinary Integration of Deutsche Airbus Specialists**.

On the highest level there exists a global model of aircraft performance and economy as a function of its specification and a set of design variables. The analysis routines in the global model are supplied by the departments who have final responsibility. In about an hour numerical optimization of these design variables will provide an estimate of the aircraft's main characteristics and geometric dimensions.

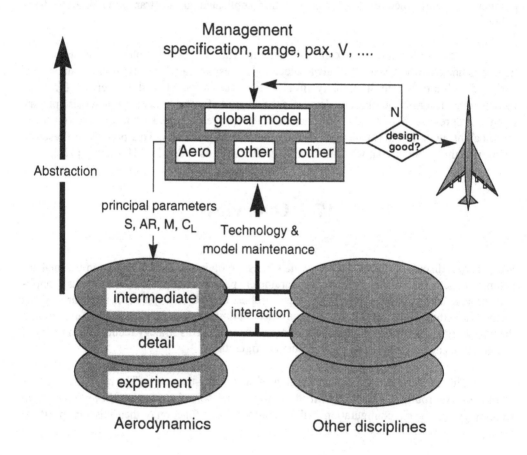

Figure 96   Overview of the MIDAS design process

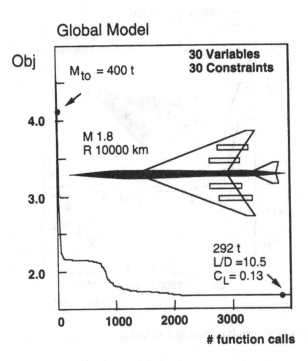

Figure 97   Global optimization

Figure 97 shows the convergence history of a global optimization.

After the optimized design is finished it is critically evaluated by the specialists. Once the decision is made to develop a design further, the departments verify the pertinent results of the global optimization in greater detail. This verification includes the design of a more detailed geometry on the basis of the global optimization output, much like putting the flesh on a skeleton. The values predicted by this intermediate model are compared to those for the global model and if necessary the global model is corrected and the process is repeated. Typically, no more than two iterations are required. In some cases the intermediate design level is also automated resulting in a closed global-intermediate loop. In this case only computational limitations prevent further integration. Note, that at the intermediate design level, within each department a new group of parameters must be optimized which justify the simplifying assumptions made at the global level. For instance in the global aerodynamic model, the assumption is made that there exists a wing camber distribution that has near minimum potential drag for a given planform. The wing planform is optimized during the global optimization cycle. At the intermediate level, the camber (represented by shape functions) is then optimized to achieve minimum lift dependent wave drag for this planform and lift. Finally the resultant ratios of the achieved minimum lift dependent wave drag to the global theoretical minimum is used as a correction factor in the next global optimization. Figure 98 shows the convergence history of the intermediate aerodynamic optimization for several optimizers.

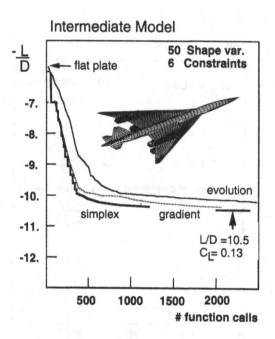

Figure 98   Intermediate aerodynamic optimization

If this ratio exceeds an allowed percentage, the model itself should be corrected. The intermediate model should not introduce parameters that conflict with the global model variables. A detailed description of the intermediate model can be found in references [371] and [373].

To date we have used this method to obtain a number of SCT configurations. The configurations presented here have many of the features of these proprietary designs, but do not present the actual configurations.

17.5.1 Present Method Issues

For successful optimization of a parameter space we have to make the objective function topography suitable for this purpose. This process is described in reference [372]. In this study we used a derivative version of the Genie optimization software [364]. In addition to general engineering optimization issues, the present method adds the following issues to the development of a global model as presented in this paper:

- **Coordination of input and output.** This requirement forces the development to be coordinated by only a few people and therefore presents the most significant "bottleneck" for industrial application. Currently we have one input coordinator for each geographic location. Changes to the model are only made based on consensus between these representatives.

- **What to model?** In principle we are always modelling cost and benefit. Clear paths should be established between the major design parameters and the cost equations. The benefits of a noise suppressor are easily defined, the number of dB suppressed at a reference condition can simply be used as an input parameter. But at what cost? Here is where the specialist is invaluable. He will calculate how much benefit to the aircraft he can guarantee as a function of various physical quantities (jet exhaust velocity...) for what cost (thrust-loss and suppressor complexity..). An accurate prediction of these relations is of course to his professional advantage since that will indirectly determine his task in the intermediate detailed design.

- **Generality versus accuracy.** Computers are bad with rules of thumb since they do not know what is behind them, but they are also fast and can solve the general equations upon which the rules of thumb are based. So therefore it is helpful to base the model on physical realities, and not to fudge it to fit one existing data set.

- **Consistency.** It is desirable to use one model to determine one parameter. If this is not possible make sure that each equation representing the same parameter contains the identical set of variables and that the value and the derivatives at the cross-over points are continuous if possible.

    The present method has a number of advantages over more traditional design methods. This method:

- **clarifies the goals** of a design project and provide a means of communication between the disciplines. Most of the time it is not very clear what the objective of an aircraft design is. By agreeing on an objective function (for instance TOC) and a number of constraints it is possible to settle interdisciplinary differences.

- **automatically debugs** the analysis routines. Small model inconsistencies are usually not noticed by the expert user because he only trusts his model in a limited range. The optimizer will exploit any weakness in the model and therefore make it more visible.

- **cleanly compares** between competing configurations. To compare aircraft they have to be analyzed with the same technologies and missions. In addition they have to be preferably analyzed with the same set of equations.

- **reduces the number of detail-design cycles.** As the experiences at DA have shown, a good baseline design will cut the number of follow-on detail cycles, thus significantly reducing the total required time.

- **shortens the design cycles.** Project management will get a fast first estimate of influence of the specification or technology on the aircraft performance and economy.

- **allows easy post optimality analysis.** Assuming that the objective function can be linearized with respect to the design variables at the (local) minimum we can verify whether we found a (local) minimum by determining the partial derivative of the objective function with respect to the design variables. They all should be zero in the case of a constrained variable or limited by an active constraint.

- **reduces the amount of data** produced. It is no longer necessary to understand a single design point by studying complicated thumb-plots. Simple graphs can be used to scan the entire design space of best configurations. Active and nearly active constraints can be monitored to indicate important performance criteria or important technologies.

- allows the **progress** that is made in the **analysis** routines to be directly translated into **an improved design**. This provides a great deal of motivation to the individual who makes a contribution to an improvement in the model.

Some of the advantages of the current system are also described in a paper by Reimers [366].

17.6 Application of the Method to Supersonic Civil Transports

17.6.1 Introduction

To design a next generation of supersonic transport a global model was developed to investigate the performance potentials of a wide range of aircraft variants.

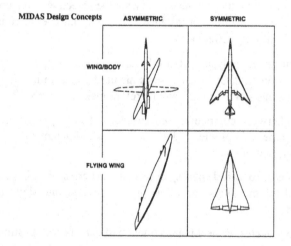

Figure 99   **MIDAS design concepts**

Figure 99 shows the three types of aircraft investigated:

a) A conventional wing-body configuration. This symmetric wing body (SWB) configuration is also the preferred configuration for military aircraft. In reference [367] Rech and Leyman provide an excellent insight into the development of Concorde, the only operational supersonic transport of this type. The SWB represents the conservative approach for a new second generation SST and all group-of-five (Boeing - MDD - NASA, DASA - AS - BAe) baseline configurations are of this type. The flying-wing as presented in Figure 99 is not a true new type since the passenger cabin is concentrated in the center of the planform and therefore not different from the conventional configuration. For these configurations the horizontal tail or canard only serves as a control surface. The horizontal tail does not contribute to the stability because of the high downwash ratio behind the low aspect ratio wing.

b) Asymmetric (oblique) wing body configuration. First proposed in 1912 by E. de Marcay and E. Moonen, this novel concept was later successfully developed by R.T. Jones at NASA Ames Research Center during 70's. A large volume of work is available on oblique wing-bodies, notably references [362] [365][358]. Recently, Elliot, Hoskins and Miller at General Electric Company [360] called for a revival of the the oblique wing program because of its larger performance margins and its low environmental impact. This concept has also an appealing passenger cabin layout.

c) Asymmetric flying wing or Oblique Flying Wing OFW. First proposed by R.T. Jones [363] of NASA Ames, the OFW was described in detail in previous references by the first author [371][368] [369][370] and by Waters, Ardema and Kroo [375]. The OFW offers a possible solution to the economic and environmental problems associated with supersonic flight due to a combination of good aerodynamic and structural characteristics. Seat modifications may be necessary to cope with a future modification of FAR 25 for an OFW with a spanwise oriented cabin. Such a cabin is desired to take advantage of span-loading. A proof of concept test flight was made with an unstable OFW at NASA Ames on May 10, 1994, by S. Morris and B. Tigner.

17.6.2 Global Model Technology Base

All aircraft are evaluated with the same analysis routines. All aircraft are designed with the same level of structural, aerodynamic and propulsion technology. This level is based on that achieved by a supersonic transports that can be produced in 2005. The most important parts are:

Aerodynamics

Skin friction is based on fully turbulent flow with characteristic sand grain roughness. It is not likely that laminar flow can bring a real advantage to the SCT before 2005. Unresolved issues are the absence of experimental facilities, the high-Re numbers, the wave drag penalty incurred due to the required low-cross flow pressure distributions and the weight- and space penalty due to the suction equipment. The inviscid flow drag is based on the theoretical minimum potential drag for a given distribution of lift and volume. This drag is then corrected based on the actual achieved drag levels in the detailed design. This correction factor therefore reflects the potential drag improvement that is still achievable. Factors of 1.1 - 1.2 for the volume dependent wave drag and the the lift dependent wave drag are typical for a high performance design. [371] describes the 1992 status of the aerodynamic global model and its corresponding intermediate model applied to the design of an ofw. Ref. [373] also shows an example of its application to an arrow wing aircraft.

Structures

Structural calculations are based on a mix of composite materials and metallic alloys. DA experience shows that intermediate carbon fibers with BMI resins achieve strain levels in excess of 0.5 % resulting in weight savings over conventional primary structures in excess of 25 %. The airframe life was specified to be over 75,000 hours with 50,000 supersonic flying hours and 25,000 pressure cycles. A minimum skin thickness of 2 mm was specified to minimize foreign object damage. An overview of SCT structures and materials technology can be found in Ermanni's publication [361]. Weight is calculated with shell-theory. [371] describes the 1992 status of the structural model. Reference [359] presents the intermediate model applied to the design of an arrow wing fuselage.

Propulsion

A turbofan engine with mixing and a variable throat area was used. Polytropic component efficiencies are typical of a future generation: between 90 and 93% for the compressor, fan and turbine. The maximum turbine entry temperature was limited to 1800 K and the maximum pressure ratio to 50. Noise suppression of up to 14 dB is allowed, but a penalty in thrust and weight is paid.

Economy - Objective Function

Since aircraft cost is used to compare the various configurations, the cost model is one of the most important. The DA economic model represents the cost structure of a european airline. The objective function that is used throughout this study is total operating cost relative to a subsonic reference. In some cases this operating cost was corrected for the expected increase in ticket price to account for the block time savings. The influence of the cost of development on sales price is included. Therefore, this study compares the relative operating cost of various aircraft for the same mission, permitting comparisons of one concept with respect to the other.

The influence on the right objective function selection on the relative "goodness" of an aircraft is clearly demonstrated by Table 3. Assume a fully loaded 250 passenger aircraft designed for a range of 9000 km with cost calculated for a reference range of 6000 km. We consider five types of aircraft 3 symmetric wing bodies designed for Mach 2. (a), 1.6 (b)and 0.85 (c) respectively and an oblique flying wing designed for Mach 2.0 (d) and 1.6 (e) and an oblique wing body designed for Mach 1.6 (f). Based on DOC and M_{to}, aircraft c (subsonic) is the best followed by e, d, b, a and f. Including IOC's and the expected increase of revenue the Oblique Flying Wings d, e are the best followed by a, b, c and f. This comparison clearly shows the necessity to agree on a common objective function.

Type	M_{to}(t)	DOC	TOC	TOC-Δ Rev
a: M 2.0 SWB	286	3.66	6.22	4.72
b: M 1.6 SWB	266	3.46	6.03	4.74
c: M 0.8 SWB	139	2.54	5.19	5.03
d: M 2.0 OFW	251	3.28	5.79	4.28
e: M 1.6 OFW	238	3.17	5.71	4.41
f: M 1.6 OWB	323	4.14	6.79	5.50

Table 3 Effect of objective on goodness

Technology Standard

Supersonic aircraft are assumed to have some technologies that are not required for subsonic aircraft:

- Flutter mode control load alleviation, which would allow increased sweep with thin wings.

- Active stability and control for all supersonic aircraft. Both the symmetric and asymmetric configuration have similar stability margins and have their neutral point at the MZFW center of gravity location.

- Powerplant variable geometry inlet and nozzle design.

- Improved navigational and environmental control systems.

To obtain a good impression about the impact of the sorts of technologies proposed for the supersonic transport they were applied to a A340 design. We took the A340 specification and optimized the configuration with the same models, requirements, technology factors and constraints as applied to the 2005 supersonic transport. Table 4 shows the results of this study. The first column shows the actual A340 data. Ref. [371] is a single point optimization of an aircraft with the A340 mission and technology[1]. These difference between the optimized and the datum A340 are well outside the spectrum of the accuracy of the global method. Ref. [373] is the same specification with the 2005 supersonic technology. The M_{to} dropped by 31 %. In the last column the cabin standard and the range are reduced to SCT specification. This SCT standard is 10 % first class (40" pitch), 30 % business (36" pitch) and the rest economy (32" pitch).

---

1. The original A340 design was not a single point design and included the requirement for a communal A330 / A340 wing

	A340-data	Ref. 1	Ref. 2	SCT
L/D	20.0	19.7	19.7	19.2
s.f.c.	0.59	0.58	0.58	0.58
DEM (t)	127	125	80	71.9
M_{to} (t)	271	275	189	155
Range (km)	13900	15000	15000	11000
Cabin	A340	A340	A340	SCT
Structure	Alu	Alu	CF	CF

Table 4 Technology Standard

17.6.3 Variables and Constraints

Input requires more than 200 variables for a full definition of the aircraft. Roughly one third of these variables are technology constants and are assumed to remain constant.

All designs were optimized with respect to the following 30 engine cycle, mission and geometry variables:

- **Maximum takeoff mass** M_{to}

- **Powerplant**. Bypass ratio BPR. Maximum turbine entry temperature SLS $T_{t,4,ref}$. The corresponding total pressure ratio ε_c. Combustor temperature cutback during takeoff $\Delta_{T4,to}$. Combustor temperature cutback at takeoff noise control point and start climb $\Delta_{T4,cb}$. The thrust margin variable allows the engines to be oversized with respect to start cruise. Noise suppression can be installed to achieve FAR 36.

- **Mission**. The flight profile is shown in Figure 100. Thrust, sweep, altitude and Mach number can be set at takeoff, takeoff thrust-cutback, midpoint climb, start cruise and end cruise, as well as at holding and diversion. The landing and approach are executed at maximum landing weight. For a specific mission segment definition point either the Mach number or the altitude was fixed during the optimization. The variable mission thrust settings make it possible to meet the noise regulations with thrust - cutback. Since the design loads are calculated based on the prescribed mission it is possible to minimize the aircraft loads in conjunction with the geometry. The lift-off speed is not less than 1.2 times the minimum control speed. In particular, the delta-wing configurations tend to take off at lift-off speeds greater than this value because of high take-off drag.

- **Dimensions**. Wing Area S. Wing Aspect ratio AR is only a design variable for wing-body configurations. For flying wings, the aspect ratio is determined by the wing area S and the wing thickness-to-chord ratio and the heuristic design rule that the payload should fit inside the structure without excess (vertical) space. Fuselage length l_f is only a design variable for wing-body configurations. The configuration fineness ratio determines the fuselage diameter of SWB's wings and the maximum absolute thickness of OFW's. Wing Root thickness-to-chord ratio t/c_{root} is an important variable to trade off wave drag against wing weight. In the case of the OFW the thickness always converged to values around 19 %. This high value can probably only be achieved by employing boundary layer control devices.

- **Sweeps schedule.** Cruise sweep Λ_{cr}. This is the same as the sweep for fixed sweep configurations. Climb sweep Λ_{climb} and take-off sweep Λ_{to} are only of significance for variable sweep configurations. The sweep of the OFW is limited to $70°$ to avoid flow separation.

- **Flap schedule.** Lift increments due to flap deflection at takeoff and landing. $\Delta C_{L,flap,to}$ and $\Delta C_{L,flap,l}$.

- **Component locations.** Spanwise location of two powerplants on half wing: y_{prop1} and y_{prop2}. Spanwise location of two fuel tanks on half wing: y_{tank1} and y_{tank2} and the spanwise widths of the tanks. The spanwise location of the main gear legs y_{gear}. The x-location of the centroids of these items is determined by balance considerations. For flying wings the relative component location did turn out to be an important design variable to achieve the benefits of span loading.

Each of these variables are constrained to the domain of their validity. In addition a large number of constraints have to be satisfied.

- **Geometry.** The powerplants (4) , gear legs (3) and fuel tanks (2) are constrained to not interfere with each other's location. The powerplants are placed outside the direct vicinity of the passenger cabin to protect passengers from engine explosion debris. The gear track was limited to 35 m to enable operation from runways that are at least 150 feet wide. The volume of the fuel tanks was sufficient to accommodate all the required fuel as well as a fuel volume equivalent to half the payload. The mean trailing edge sweepback angle was limited to $30°$ for arrow wings due to the expected aero-elastic and control problems of more highly swept symmetric thin wings [356]. The maximum aircraft width and length is constrained to the current large aircraft gates at Frankfurt airport. This leads to a maximum structural span of about 120 m for OFW's.

- **Performance.** The range was constrained to a value not smaller than the design range. The takeoff field length and landing field length was constrained to 3300 m (11000ft) to allow operations from most international runways. The one-engine out screen height climb requirement was set in accordance with the FAR25 regulations. A minimum all engine climb gradient of 4% must be maintained at the takeoff FAR-36 cutback thrust level. The approach angle is $4°$.

- **Environmental constraints.** All the designs in the study meet the FAR 36 stage 3 levels with tradeoffs as formulated in ICAO an. 16 3.5. No sonic boom or ozone depletion constraints were imposed. Sonic boom can not be avoided, the aircraft is therefore constrained to flying over water. No adverse effects on sea mammals due to overwater flights have been recorded.

The width of the sonic boom carpet increases with Mach number, and therefore it will be
much easier to use sonic boom corridors at the lower Mach numbers. Such a trip could involve going from Munich across the Adriatic sea (M1.6 supersonic) to the Sinai Peninsula
(M 1.15 boomless) through the red sea (M1.6) to India. At this time there are no models available to accurately predict the effects of aircraft effluents on the ozone layer.

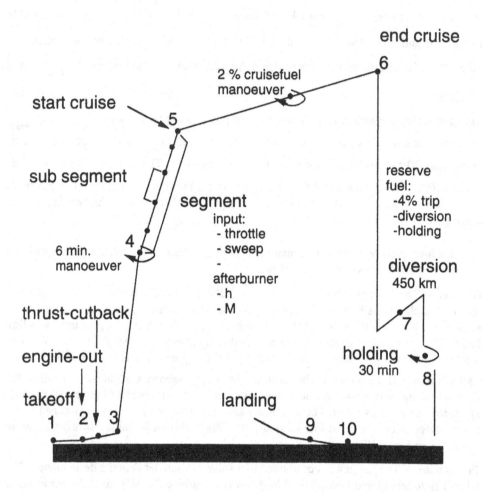

Figure 100   Flight profile

17.7 The Space of Optimized SCT's

In this section we will compare optimized configurations. The comparisons are made for designs with fixed mission parameters, such as constant range, payload, and Mach number.

17.7.1 Reference Aircraft

The baseline mission is 250 passengers over a 9000 km range. For the reference subsonic transport with a cruise Mach number of 0.8 Table 5 shows a cruise lift-to drag ratio of about 18, the same can be said for the specific fuel consumption of 0.61. The wing has a typical dimensions for a conventional transport. Also the weight fractions are very much like those of todays aircraft.

A reference Mach number of 1.6 is chosen so the design would to easily meet future ozone layer depletion regulations and sonic boom corridor limitations. Choosing a higher Mach number would also make range, payload and Mach number variations difficult due to the high number of active constraints. At Mach 1.6 the OFW is clearly superior to the SWB. Interestingly enough, the cruise lift-to-drag ratio of the OFW is close to that of the SWB. The optimizer chose to improve the lift-to-drag of the SWB up to the level of the OFW at the expense of the structure. The other explanation is that the size of the OFW could be reduced by increasing the thickness-to-chord ratio and that this outweighed the decrease in lift-to-drag. The absolute drag the OFW is still 16 % less than that of the SWB ! The structure of the OFW is also 14 % lighter than for the SWB. In the case of the OFW, the lift-to-drag ratios of around 10.5 during cruise were achieved without a large structural penalty because of span loading and the fact that the OFW wing is much thicker. The OFW has similar TOC's as the subsonic transport. Though both configurations are not constrained by sonic boom or ozone layer depletion, they are both constrained by takeoff and sideline noise. Because of its better takeoff lift-to-drag ratio and lower wing loading the OFW cuts back the engine at takeoff to meet the noise requirement. The SWB uses most of the suppression that is allowed. The price for suppression seems to be less than the price for the extra installation drag of a higher bypass ratio. The OFW mean sonic boom overpressure is only half of that of the SWB. The OWB is not discussed as a reference aircraft because of its poor performance.

17.7.2 Variation of Range, Mach number and Payload

Range

Figure 101 shows that this OFW is able to achieve ranges of up to 6500 nm (12000 km). Spanloading enables the OFW to improve both range and TOC. The OFW has has a similar reduction of TOC with range as the subsonic reference. The trip cost due aircraft ground handling, baggage handling, administration and landing fees are only weakly dependent on range. As a consequence the indirect operating costs (IOC) per available seat km reduce with range. This effect is not so clearly visible for the SWB's. Beyond 6000 km range the direct operating cost per available seat increases fast. And at 9000 km it increases faster than the decrease of the IOC's. Relative to the subsonic reference the SWB seems to be most attractive for the transatlantic range. Beyond 11000 km its maximum takeoff weight snowballs, even with our aggressive technology assumptions. The OWB even has problems making it across the pacific.

Type	SWB	SWB	OFW
Cruise Speed	M0.8	M1.6	M1.6
Wing Area (m^2)	304	625	1290
Root,Tip t/c (%)	15/11	4.9/3.5	19/13
l.e. Sweep	29 deg	57 deg	70 deg
cabin l x w (m)	44 x 4.9	55 x 4.2	25 x 8.0
total l x w (m)	62 x 49	89 x 47	120 x 15
Weights:			
DEM (t)	61	114	99
M_{to} (kg)	139	266	238
Powerplant:			
SLS Thrust (kN)	4 x 63	4 x 196	4 x 210
BPR	5.1	0.7	0.8
$\varepsilon_{c,max}$	50	44	32
$TT4_{max}$ (K)	1800	1800	1800
Operation:	Initial Cruise		
L / D	17.9	9.9	10.5
s.f.c.	0.61	0.94	0.96
h	11200 m	13300	14800
TOC *ct/seat.km*	ref.	+16%	+10%
Environment:			
Mean ΔP ($N \over m^2$)		95	44
SL Noise (dB)	89	103	102
TO Noise (dB)	102	106	103
AP Noise (dB)	98	99	103

Table 5 Reference Aircraft (9000 km / 250 pax)

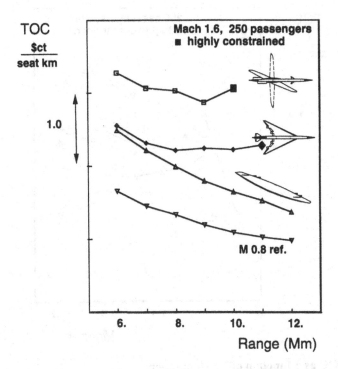

Figure 101   TOC as a function of range

Mach number

Figure 102 shows the influence of the increase in speed from Mach 0.8 to 2.0 on TOC for the SWB's and the OFW's. The dominant influence is the price of wave drag added to supersonic configurations. At supersonic speeds the increased specific airframe complexity increases the purchase price and therefore the direct operating costs. Over a wide range of Mach numbers the SWB's and the OFW's have similar lift-to-drag ratios and similar specific fuel consumptions. The structural weights are completely different. Both concepts have their minimum TOC at Mach 1.6. Mach numbers in excess of 2.0 are probably not possible due to thermal stability problems with conventional fuels for trips in excess of 3 hours. In addition we found that a number of these designs were limited by as many as a dozen constraints at the same time. This is caused by incompatibility of low-speed and high-speed flight. The OWB is worse than the SWB except at very low supersonic speeds. As we discussed before, the OWB improves more than the SWB if we consider a lower range. Our calculations therefore agree with the 1977 Boeing High-Transonic Speed study by Kulfan, Neumann et al. [365] that stated that the OWB is slightly superior to the SWB at Mach 1.2 and a 5500 km range. Our calculations however, do not support the claims made by the 1991 GE study of Elliot, Hoskins and Miller [360] that this configuration is superior to the SWB at 9000 km and Mach 2.4. At Mach 1.8 the lift-to-drag of this configuration is already down to 7.5 because of excessive wave drag. In addition the structure is very poor because of the single pivot that has to transfer all the loads.

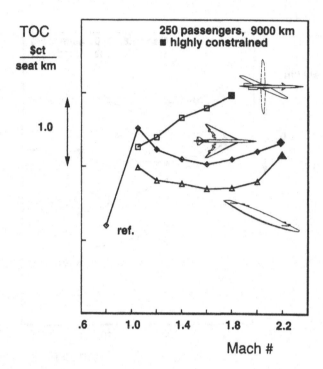

Figure 102   TOC as a function of Mach number

Payload

The overall trend that the TOC's reduce with increased payload is caused by the reduction of the DOC's per passenger km. with payload. Maintenance costs, fuel burn and depreciation do not rise (exactly) proportionally to the number of passengers transported. This effect is even a bit stronger for the supersonic transports. The larger area-ruled supersonic transports have a better cabin volume efficiency then the smaller ones. However, only supersonic configurations that are not highly constrained benefit from an increase in payload with respect to the subsonic transports. Because of the poor volume efficiency of the OFW configurations, the standard payload of 250 passengers may not be large enough. For the OFW a payload of 400 passengers is better. It is not possible at this time to design a 400 passenger SWB SCT configuration since such an aircraft would have an excessive M_{to}. However this study does seem to justify the approach of the American HSCT teams (NASA, Boeing, MDD) [357] to transport the most passengers for a given M_{to}. The M_{to} is for all practical purposes limited to 350 tons. At higher takeoff weights the critical sideline noise constraint is no longer relaxed to accommodate heavier aircraft. Figure 103 shows the effect of increased payload on the TOC of the OFW. If we compare the 250 and 400 passenger OFW's we find that the maximum takeoff weight has increased 33% while the payload has increased 60%. A detailed explanation of the improved economy of the oblique flying wing transport over the symmetric wing body is published in ref. [371].

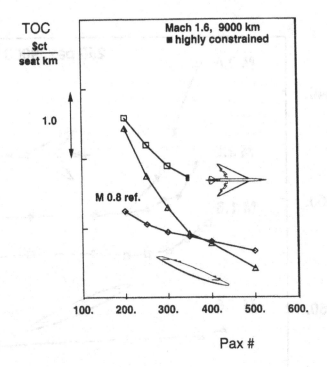

Figure 103   TOC as a function payload size

17.7.3 Technology Assessment

Evaluating the influence of new technologies on the aircraft design and get the benefits (or penalties) in the operating cost is a major part of the work in the future project office. Such studies were traditionally done by fixing all parameters and make a one dimensional analysis. But this approach does not allow for second order effects which can compensate or enhance the influence of the parameter on the aircraft design. The MIDAS concept enables the project engineer to evaluate the consequences of technology introduction fast and considering all aspects of the global model.

For the supersonic transport of the next generation an optimistic maximum noise suppression level of approximately 14 dB must be taken into account for reaching the goal of FAR stage 3. The weight increase and the thrust loss due to the noise suppressor are incorporated in the model as [$10^{-0.083}\Delta dB$] % loss of thrust and 5% increase in nacelle weight per dB of suppression. Figure 104 shows the change of M_{to}, if the technical target of the year 2005 will not be achieved or surpassed. The OFW shows a relatively neutral behavior, due to his flexibility to compensate surpressor nonperformance by reducing thrust requirements. For the SWB a shortfall of the target will be fatal especially for Mach numbers over 2.0.

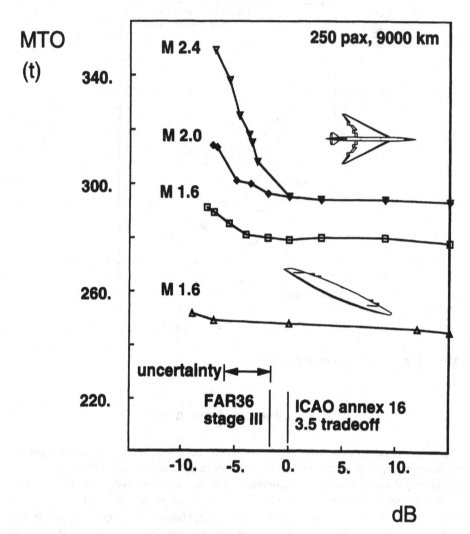

Figure 104   Effect of noise regulations

In ref. [374] further examples of technology assessment studies with MIDAS are given.

Acknowledgements

The author would like to thank all the people at Deutsche Airbus SCT-team who contributed to make this project especially our project manager Detlev Reimers and Dirk von Reith in Hamburg. Special thanks also to Prof. Kroo's aircraft design group at Stanford University for their work on the Genie optimization shell and to Prof. Yoda for her constructive editing.

17.8 Conclusion

In this chapter we discussed a design strategy which given a mission finds the optimum aircraft design in the analysis parameter space. This parameter space is determined by the various disciplinary groups. Based upon this input robust physically correct analysis modules were developed. Improved non-linear optimization techniques were used to find the best design using these modules in this parameter space. This strategy reduced the number of design cycles and allowed us to evaluate more configurations. The ability to evaluate more configurations is essential for projects with high development cost that depend on the realization of aggressive technology targets. In the case of the supersonic transport this ability allowed us:

- to investigate a wide range of solutions to find one which will be flexible enough to compete in an unknown market 30 years from now.

- to build scenarios to find out what happens when the design requirements will change or technical targets cannot be reached

- to find out where we have to invest work in detailed technology programs to get the best results.

The present method was used to study the relative performance of symmetric wing body (SWB) and oblique flying wing (OFW) and oblique wing body (OWB) supersonic transports over a wide range of missions. The conclusions of this study where:

- Supersonic transports achieve their highest profit potential at Mach numbers between 1.6 and 2.0 and are able to compete with subsonic transports when adequate structural and noise reduction technology is available.

- Noise regulations and runway loads may limit the size of symmetric wing body configurations with cruise speeds over Mach 2.0 to 350 tons. Only with very aggressive technology assumptions will such a transport be able to transport 250 passengers across the pacific.

- It is possible to reduce the total operating cost per seat of supersonic tranpsports by increasing the payload size up the maximum that is allowed by runway load and noise constraints. Larger payloads use the available volume more efficiently for a given passenger cabin standard.

- The best oblique flying wings are large and long range. They are compatible to the current traffic infrastructure, and not dependent on very aggressive technology assumptions. In addition, they produce less sonic boom and can comply with stage 4 noise regulations.

- The oblique wing body is unsuitable for speeds over Mach 1.4, payloads over 200 passengers and ranges in excess of 6000 km. Both its aerodynamics and structures are too poor for higher transport performance.

17.9 References

[356] **Bobbitt, P. J.**
Theoretical and Experimental Pressure Distributions for a 71.2 Degree Swept Arrow-Wing Configuration at Subsonic, Transonic, and Supersonic Speeds, NASA Langley Res. Center NASA CP-001, 1976

[357] **Boeing Commercial Airplane Company**
High-Speed Civil Transport Study, NASA Contractor Report 4233, September 1989.

[358] **Boeing Commercial Airplane Company**
Oblique Wing Transport Configuration Development, NASA CR-151928, 1977

[359] **Dittert, M., Ermanni, P., Albus, J.**
Auslegung und Bewertung von Strukturkonzepten für den Rumpf eines zukünftigen Überschallflugzeuges (SCT), Externe Studienarbeit Institut für Leichtbau, RWTH Aachen, Nov. 1993

[360] **Elliot, D.W., Hoskins, P.D., Miller, R.F.**
A Variable Geometry HSCT, AIAA Aircraft Design Systems and Operations Meeting, September 1991 Baltimore, paper AIAA 91-3101. 1991

[361] **Ermanni, P.,**
Assessment of a future SCT with regards to structures and materials, The Supersonic Transport of the Second Generation, 7th European Aerospace Conference EAC '94. 25-27 October 1994 Toulouse.

[362] **Graham, L. A., Jones, R. T., Boltz, F. W.**
An Experimental Investigation of Three Oblique-Wing and Body Combinations at Mach Numbers Between 0.60 and 1.4, NASA TM-X-62,256, 1973

[363] **Jones, R.T.**
The Supersonic Flying Wing, Aerospace America, November 1986

[364] **Kroo, I.**
An Interactive System for Aircraft Design and Optimization, Aerospace Design Conference, AIAA 92-1190, 1992

[365] **Kulfan, R. M., Neumann, F., et al.**
High Transonic Speed Transport Aircraft Study, NASA CR-114658, 1973

[366] **Reimers, H. D.**
Das Überschallverkehrsflugzeug der 2. Generation - Eine Zweite Chance, Paper DGLR Jahrestagung 1993, Sept 1993

[367] **Rech, J., Leyman, C. S.**
A Case Study by Aerospatiale and British Aerospace on the Concorde, AIAA Professional Study Series, 1979

[368] **Van der Velden, A. J. M.**

The Conceptual Design of a Mach 2.0 Oblique Flying Wing Supersonic Transport, NASA CR 177529, May 1989

[369] **Van der Velden, A. J. M., Kroo, I.**
A Numerical Method for Relating Two-and Three-Dimensional Pressure Distributions on Transonic Wings, AIAA/AHS/ASEE Aircraft Design and Operations Meeting, AIAA 90-3211, 1990

[370] **Van der Velden, A. J. M., Kroo, I.**
The Sonic Boom of an Oblique Flying Wing, *Journal of Aircraft* jan-feb 1994.

[371] **Van der Velden, A. J. M.**
Aerodynamic Design and Synthesis of the Oblique Flying Wing Supersonic Transport, PhD-thesis Stanford University, Dept. Aero Astro SUDAAR 621, Univ. Microfilms no. DA9234183, June 1992

[372] **Van der Velden, A. J. M.**
Tools for Applied Engineering Optimization, VKI lecture series in Optimum Design Methods in Aerodynamics AGARD R 803, April 1994.

[373] **Van der Velden, A. J. M.**
Aerodynamic Shape Optimization VKI lecture series in Optimum Design Methods in Aerodynamics AGARD R 803, april 1994.

[374] **Van der Velden, A. J. M., Von Reith, D.**
Multi-Disciplinary SCT Design at Deutsche Aerospace Airbus , 7th European Aerospace Conference EAC '94. 25-27 October 1994 Toulouse.

[375] **Waters, M., Ardema, M., Kroo, I.**
Structural and Aerodynamic Considerations for an Oblique All-Wing Aircraft, Aircraft Design, Systems and Operations Conference AIAA 92-4420, Hilton Head August 1992.

CHAPTER 18

LAMINAR FLOW FOR SUPERSONIC TRANSPORTS

J. Mertens

Daimler-Benz Aerospace Airbus GmbH, Bremen, Germany

18.1 Introduction

Supersonic transports are very drag sensitive. Technology to reduce drag by application of laminar flow, therefore, will be important; it is a prerequisite to achieve very long range capability. In earlier studies it was assumed that SCTs would only become possible by application of laminar flow [376]. But today, we request an SCT to be viable without application of laminar flow in order to maintain its competitiveness when laminar flow becomes available for subsonic and supersonic transports. By reducing fuel burned, laminar flow drag reduction reduces size and weight of the aircraft, or increases range capability -whereas otherwise size and weight would grow towards infinity. Transition mechanisms from laminar to turbulent state of the boundary layer flow (ALT, CFI, TSI) function as for transonic transports, but at more severe conditions: higher sweep angles, cooled surfaces; higher mode instabilities (HMI) must at least be taken into account, although they may not become important below Mach 3. Hitherto there is a worldwide lack of ground test facilities to investigate TSI at the expected cruise Mach numbers between 1.6 and 2.4; in Stuttgart, Germany one such facility -a Ludwieg tube- is still in the validation phase. A quiet Ludwieg tunnel could be a favourable choice for Europe. But it will require a new approach in designing aircraft which includes improved theoretical predictions, usage of classical wind tunnels for turbulent flow and flight tests for validation.

18.2 Future Supersonic Commercial Transports

Realisation of a new Supersonic Commercial Transport (SCT) must meet challenging environmental and cost requirements (Figure 105). It can be certified only when meeting the future rules on emissions and noise, and if it will bring some profit to manufacturers and airlines. Drag is directly related to fuel burned, emissions and operating costs; but it influences also noise via aircraft size and weight. At the time being, the perspective for a new SCT improves; although emission restrictions may question "if", whereas noise and costs ask "when" it becomes possible.

Figure 105   A New Supersonic Commercial Transport

An SCT makes sense only for long ranges of at least 2000 nm. The longer the distances the more it becomes attractive for the passengers. At flight times of more than 4 to 6 hours most passengers feel uncomfortable and many see flight time as a waste of time, even tourists. But the SCT has to compete against future new subsonic aircraft (Figure 106) providing more space and comfort for better accomodation of a long flight time. These aircraft will have low operating costs which cannot be met by an SCT. So the SCT has to compete with speed comfort and productivity against efficiency and space comfort.

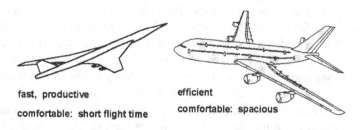

Figure 106   Future Long Range Aircraft

To meet these challenges, the aircraft must be optimized mainly respecting the four different design points:

- very efficient supersonic cruise at Mach 2 to 2.4,

- quiet take off and landing at steep flight path angles,

- high subsonic cruise capability (Mach 0.9) for flight over inhabited areas, where supersonic flight is not permitted,

- transonic acceleration at about Mach 1.1.

For all four points minimisation of aerodynamic drag is one of the challenges. Drag contributions are:

- wave drag

- vortex drag

- friction drag.

In this chapter emphasis is put on friction drag reduction by laminarisation.

An SCT has a large wing area, about 3 times the size of comparable subsonic aircraft (Figure 107). It cruises at low angle of attack and low lift coefficient C_L due to the high lift dependent drag (consisting of wave drag and induced drag). Friction drag contributes by about 35% to overall drag during supersonic cruise. It can significantly be reduced by laminarisation. Present SCT-designs achieve a lift to drag ratio (L/D) of about 8.5 for a turbulent wing. With partial laminarisation nearly 10 may be possible. The goal is at 9.5 for turbulent and 11 for laminar flow. The maximum range of a 250 - 300 passenger SCT in the year 2010 is expected with a turbulent wing at less than 5000 nm and with laminarisation at about 6000 nm, for a take off weight below 400 tons.

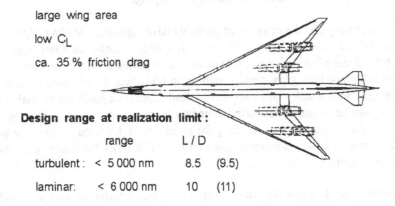

large wing area

low C_L

ca. 35 % friction drag

Design range at realization limit :

	range	L / D	
turbulent :	< 5 000 nm	8.5	(9.5)
laminar:	< 6 000 nm	10	(11)

Figure 107   SCT-Charateristics

18.3 Laminarisation of Supersonic Transports

Efficient laminarisation of supersonic transports requires a hybrid approach. Figure 108 shows a concept oriented on the different physical properties:

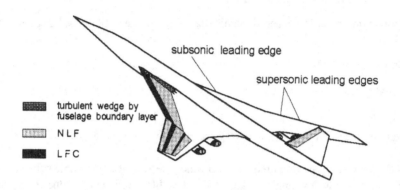

Figure 108   Possible Laminarisation Scheme

The inner wing has high sweep angle with subsonic leading edges (the leading edge stays within its own Mach cone). Therefore the leading edge is rounded with a radius which still inhibits usage of pure Natural Laminar Flow (NLF) techniques, but requires Laminar Flow Control (LFC). NLF means that laminarisation is achieved solely by suited shaping which controls pressure gradient and - at the leading edge - 3D divergent flow. LFC uses artificial measures to alter the boundary layer flow, e.g. boundary layer suction, to improve the boundary layer velocity profiles with respect to laminar disturbance damping.

The outer wing and the empennage probably have supersonic leading edges (the leading edge is outside its own Mach cone). Those leading edges usually are sharp. Possibly NLF can be used behind sharp leading edges, at least as long as profile curvature remains very small. But local Reynolds numbers grow rapidly leading to transition. In addition, curvature introduces pressure gradients which - on swept wings - lead to pressure gradients normal to the flow direction; this provokes some destabilizing boundary layer crossflow waves. Suction can prolong the laminar flow region. Before reaching the hinge line of deflected rudders, the boundary layer should become turbulent in order to avoid unfavourable shock/boundary layer interferencies with laminar separation.

As long as the fuselage boundary layer is turbulent, a turbulent wedge extends at the wing along the intersection with the fuselage. This is a significant part of the surface of SCT - wings which have large root chord length.

In a NASA-study [377] Boeing has investigated the impact of laminarisation on supersonic transports. Table 1 shows the improvements due to laminarization for a 250 passenger transport, design Mach number 2.4 and optimistic - but at least comparable - design ranges of 5000 nm and 6500 nm. The additional weight for the suction system of about 4.5 tons and the thrust reduction by 0.2% are negligeable when looking at the benefits:

- reduced fuel heating, essential at Mach 2.4 over long ranges

- important weight reductions

- significantly reduced fuel burned.

This was demonstrated in a first study neglecting some snow ball effects; further improvements after optimisation are expected.

Suction System		**Benefits for**	5 000 nm	6 000 nm	range
weight	4,5 t	fuel heating	-25.0 %		
thrust reduction	0,2 %	MTOW	-8.5 %	-12.5 %	
		OEW	-6.0 %	-10.0 %	
		block fuel	-12.0 %	-16.0 %	
		MTOW turb.	350.0 t	530.0 t	

MTOW: maximum take-off weight   OEW: operational empty weight

Table 1: Laminar Flow SCT-study [377]- Mach 2.4, 250 passengers

The 6500 nm turbulent aircraft could not be realized with the assumed weight limit of 500 tons. But according to our experience, probably the weight of an SCT must even stay below 400 tons for take-off noise limitations. Over 400 tons all designs seem to diverge; especially because the absolute noise limit is reached, whereas at lower weight noise limits are related to take-off weight.

18.4 Supersonic Transition Physics

At the subsonic leading edges of the inner part of the wing, Attachment Line Transition (ALT) occurs as known from subsonic transports (Figure 109). It develops because on swept wings the flow at the attachment line does not start its contact with the surface there (with local Reynolds number zero), but follows the attachment line and splits the flow to the upper and lower side of the wing. For infinite swept wings a boundary-layer develops at the attachment line which is in an equilibrium between boundary layer material advected along the attachment line (increasing boundary-layer mass) and divergent flow by the removal of mass over the wing (reducing boundary-layer mass).

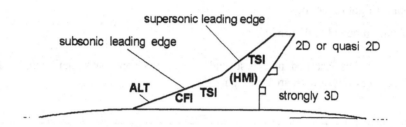

Figure 109   Transition Types

Downstream the round leading edge a strongly three dimensional flow region produces Cross Flow Instabilities (CFI), stronger than on transonic aircraft with moderately swept wings. Usually, CFI waves are the "stationary" vortical waves in the boundary layer, with the wave front direction along the stream line or the wave normal perpendicular to the stream line. Because they are oriented on the streamline and do not move, the disturbances accumulate along the stream line.

Further downstream, wing geometry presents a large region with very low surface curvature. Here nearly conical flow conditions prevail. Typical two-dimensional disturbances are the Tollmien-Schlichting-Instability waves (TSI). At low speeds, TSI-waves have their wave front normal to the flow direction, but they move in flow direction. In supersonic flow, TSI waves are inclined to the flow direction, so that the wave front direction is between normal and Mach angle, close to the Mach angle of the inviscid flow. In addition, instability waves of other orientations and wave speeds occur, but CFI (with nearly stream line orientation) and TSI usually are the most important.

On SCTs the conical directions of nearly constant flow conditions do not coincide with the main flow direction, nor are they orthogonal, but strongly inclined. This gives even in the nearly conical region on the wing 3D flow influences tending to a build-up of vortices. So, although Tollmien-Schlichting Instabilities (TSI) develop, CFI remain valid here. Strong interaction must be expected.

Behind the *sharp* supersonic leading edges of the outer wing and empennage, the flow is only two-dimensional, if there is a flat surface behind the leading edge (wedge flow). ALT does not exist, TSI develops, but as soon as there is curvature, CFI becomes important [378].

In supersonic flow, in addition to CFI and TSI, Higher Mode Instabilities (HMI) may occur: These are waves travelling at supersonic speed relative to the undisturbed flow. For flat plates these HMI occur only at free stream Mach numbers above 3 [379]. It is not expected that they become important for supersonic transport; although, at first, they must not be neglected for the more complicated 3D-flows.

TSI and HMI are sensitive to changes in the boundary layer temperature profile: TSI are damped by cooling (i.e. surface temperature lower than recovery temperature of the air flow), whereas HMI are amplified by cooling; for heating vice versa. All surfaces on a supersonic transport being of any interest for laminarisation do more or less cool the boundary layer flow. Two cooling mechanisms are important: capacity cooling by the heat sink of structure and fuel, and radiation cooling due to the elevated surface temperature. The latter has no big influence at the relatively low temperatures of supersonic transports (mostly less than 450 K), but it prevents assumptions of adiabatic flow. Heating surfaces only occur during deceleration periods or at the engines.

18.5 Theoretical Prediction of Supersonic Transition

ALT is predicted by the Pfenninger/Poll criterion [380],[381] which is assumed to be valid at supersonic speeds (Table 2); but confirmation at the Mach numbers of interest is missing. Means to avoid ALT can be transposed from subsonic knowledge. Additional investigations are still strongly appreciated: because it is still a big challenge to design and manufacture a subsonic leading edge for high aerodynamic performance at cruise and take-off which avoids ALT and is able to control CFI by suction.

ALT: Attachment Line Transition   CFI: Cross Flow Instability

ALT:	Pfenninger / Poll criterion, as for subsonics
CFI, TSI:	+ strongly 3D + coupling of CFI and TSI + temperature profile (cooling) + higher mode instabilities (HMI)
Required:	+ improved linearized theory ("e^N") + improved analysis (PSE, DNS) + **accurate 3D-solution of undisturbed flow**

TSI: Tollmien-Schlichting-Instability   HMI: Higher Mode Instabilities

Table 2: Transition Effects and Prediction

CFI, TSI and HMI are predicted by stability analysis of the boundary-layer flow disturbances.

Linearized theory ("e^N") has matured and can be used routinely, even for supersonic investigations [382], if cooled surfaces and HMI are respected. Linearized theory solves the flow equations by superposition of two parts [383]:

- undisturbed flow,
- small disturbance of one disturbance mode,
  (one frequency resp. one wave length at one inclination to the flow direction).

Because the disturbances are small, linearisation is allowed. Eventually, it is possible to derive a pure local disturbance equation formulation, where at one position the disturbance equations are solved only in normal position to the wall. Resulting is the local amplification rate for the selected disturbance mode at this position. Total amplification rate is achieved by the following procedure:

First, search for a point of indifference, where amplification rate is zero, i.e. it changes from damping to amplification. Starting from this point, follow a suited integration line (downstream), e.g. a stream line (of the "inviscid" flow). Integrate along this (stream) line the local amplification rates for the selected disturbance mode (which may change from point to point, depending on the selected integration strategy). When writing the integrated amplification rate A in exponential form

$$A = e^N,$$

$N = ln(A)$ becomes the so called N-factor. So, the result of linearized theory is an amplification rate, but not the disturbance itself. Therefore, transition prediction requires a validation by transition tests in order to calibrate a relevant amplification rate. This calibration may depend on the environment (free flight, wind tunnel, external disturbances like turbulence, noise; internal disturbances like roughness, waviness, surface vibrations). The often cited limit N-factor of about 10 is restricted to specific calculation methods (incl. the selected boundary-layer codes) and applications; (some people suggest, that 10 was only detected, because we have 10 fingers...)

By definition, linearized theory *cannot* calculate

- changes of the undisturbed flow introduced by the disturbances
- interferencies of different amplification modes
- sensitivity of transition to external or internal disturbances (so-called receptivity).

The limitations of linearized theory become increasingly obvious (Figure 110): More or less strong coupling of CFI and TSI may occur. The correlation figures for both cases are completely different and, maybe, the whole zone in between can become valid. Additional effects like curvature and boundary layer divergence must be respected; hitherto especially the latter is not taken into account and may be responsible for some confusing results. Also, validity of linearized theory for CFI is questioned [384], especially for strong CFI like on SCT-wings. In supersonic flow, the TSI-waves are not normal to the flow direction, but inclined as nearly by the Mach angle. So coupling between TSI and CFI may increase.

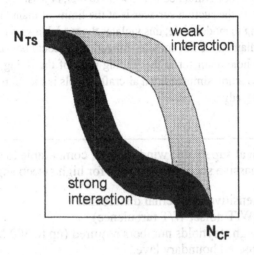

Figure 110 N_{TS}/N_{CF} for Transition Prediction

Furthermore, linearized theory calculates only amplification rates (N-factors) which require validation by experiments, but ground test facilities for supersonic transition tests still do not exist.

A remedy is seen in Parabolized Stability Equations (PSE) [385] which are able to handle coupling of CFI and TSI and - perhaps - to investigate the receptivity problem [386], [387]. The latter must be solved to understand the supersonic wind tunnel simulation problem.

Analytical approaches were limited to very special problems.

For insight in the complicated flow physics and for calibration of the simplified methods further investigations with Direct Numerical Simulation (DNS) - see e.g. [388] - is required.

For all engineering methods (linearized or parabolized disturbance theories) a prerequisite is the accurate solution of the undisturbed flow. It requires solution of the boundary layer profile for velocities and temperatures in both directions, accurate in the second derivatives. This is, for the years to come, the most severe task on strongly 3D flows.

18.6 Wind Tunnels for Supersonic Laminar Flow

ALT and CFI are not very sensitive to disturbances of the incoming flow or noise radiation (table 3). So, classical wind tunnels should be suitable for investigations. But at the high sweep angles of subsonic leading edges massive suction is required (at high model Reynolds numbers up to 300 Mio). Suction flow cannot be simulated in the wind tunnel, at least not for complete aircraft models: The hole diameter in the suction surfaces is at the limits of manufacture, hole diameter cannot be reduced according to model scale; this violates the model laws. This violation becomes important, when the hole diameter is not small compared to the local boundary layer thickness. For SCT-applications, the hole diameter at the leading edge of the flying aircraft is about the boundary layer thickness. Suction simulation on aircraft models is therefore impossible, at least in the vicinity of the leading edge.

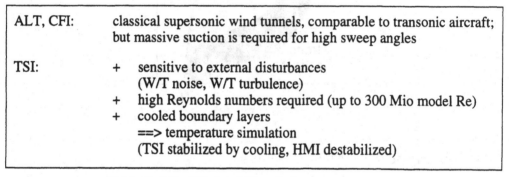

ALT, CFI:	classical supersonic wind tunnels, comparable to transonic aircraft; but massive suction is required for high sweep angles
TSI:	+ sensitive to external disturbances (W/T noise, W/T turbulence) + high Reynolds numbers required (up to 300 Mio model Re) + cooled boundary layers ==> temperature simulation (TSI stabilized by cooling, HMI destabilized)

Table 3: Wind Tunnel Simulation

TSI are very sensitive to external disturbances. For supersonic wind tunnels these are the turbulence of the incoming flow (as for subsonic wind tunnels). Also, strong noise is radiated into the test section. It is - by one part - produced by upstream noise radiated via the reservoir section, e.g. valve noise in blow down tunnels. But the most severe part is boundary layer noise radiated by the turbulent boundary layer of the wind tunnel nozzle into the test section: Each turbulent eddy in the outer boundary layer produces a small shock wave on its back which radiates a strong noise in Mach line direction. This provokes premature transition, so that effectively transition in supersonic wind tunnels seams to be dependent on nozzle Reynolds number instead of the model Reynolds number, the so called unit Reynolds number effect.

Furthermore, in most supersonic wind tunnels the attainable Reynolds numbers are completely insufficient. Often they are so low, that after provoked transition (tripping) relaminarisation occurs [389]. The cruise Reynolds numbers for supersonic transports are about $Re_L =$ 300 Millions with respect to the aircraft length L!

In the past, surface temperature of supersonic wind tunnel models was not taken into account. For investigation of TSI (and HMI), accurate simulation of the temperature profile in the boundary layer is necessary, i.e. the ratio of model wall temperature to stagnation temperature must be simulated.

To enable supersonic transition measurements, a quiet supersonic wind tunnel was developed at NASA-Langley (Figure 111) [390]. It is a small pilot tunnel for Mach 3.5:
In the subsonic part of the nozzle throat the boundary layer is removed to provide a young laminar boundary layer in the wind tunnel nozzle. This laminar boundary layer does not radiate significant noise into the test section. When the nozzle boundary layer becomes turbulent, noise is radiated. But in supersonic flow this noise follows characteristics (Mach lines). So a quiet test zone is provided, beginning with the parallel flow section and ending with the characteristics of the nozzle transition zone. This wind tunnel provided transition measurement results comparable to flight tests.

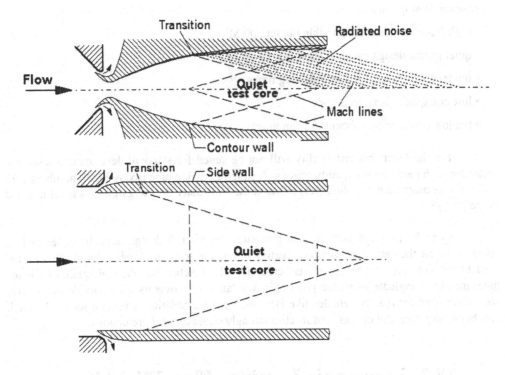

Figure 111   Quiet Supersonic Wind Tunnel

Another wind tunnel provided test data not showing the unit Re-effect: this was the Ludwieg tube in Göttingen [391] with measurements at Mach 5 (Figure 112) [392]. The reason for these high quality measurements is not completely understood: The Ludwieg tube provides an incoming flow of extremely low turbulence, but the wind tunnel nozzle has a turbulent boundary layer of high nozzle Reynolds number, i.e. with small boundary layer thickness and - perhaps - not so disturbing noise levels and spectra.

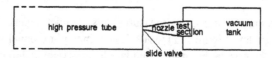

no unit Re - effect found for the Göttingen Ludwieg tube (M = 5)

Figure 112   Ludwieg Tube

With respect to future supersonic laminar flow experiments, the Ludwieg principle should be considered in Europe, possibly with a quiet nozzle - like in the US, where Ludwieg tubes are designed for SCT-tests. Some advantages are obvious:

- superb flow quality,

- high Reynolds numbers possible (about 300 Mio),

- quiet nozzle design easier at short testing times,

- model cooling easier for short testing times,

- low costs, affordable in Europe,

- but low productivity (about 0.5 s run time).

For the latter this test facility will not be suited for standard development tests, but rather for high performance quality checks. A suited Ludwieg tube (today Mach numbers 1.75 and 2.5; size more than 1 m diameter) exists at the University of Stuttgart and was refurbished in the past years.

Considering these facts, a new procedure for aircraft design must be developed. It relies more on theoretical predictions, partial simulation tests in classical wind tunnels (i.e. model tests with turbulent flow) and very carefully selected checks in the high quality Ludwieg tube, mostly to evaluate transition physics in validation experiments. Comparable procedures were developed for reentry vehicles like Hermes: Partial simulation is used, a relatively small number of experimental checks, and it relies strongly on theoretical predictions.

18.7 Supersonic Laminar Flow Flight Tests

At NASA-Langley an F-16XL was modified for supersonic laminar flow investigations (F-16XL SSLFC, Figure 113). A glove with suction was designed [393] and applied to the aircraft. It was flown with successful laminarisation by suction behind a round subsonic leading edge at supersonic speeds [394]. But apparently laminar flow was realized only in dive (at no lift) conditions, and there are difficulties in understanding the results and in validating the methods used.

The reason seems to be insufficient resolution of the boundary layer profiles in the solution of the undisturbed flow, including the second derivatives (ca. 20 points in the boundary layer normal to the surface instead of about 80 required). Nevertheless, a suction system able to provide laminar flow at supersonic flight was demonstrated.

Figure 113   F-16XL SSLC Flight Test

Since then a second flight test series with the F-16XL SSLFC was planned using two different new glove designs for the right and left hand wing (Figure 114) [395] and completed [396].

Figure 114   F-16XL SSLFC with Two Gloves

In any case, flight experiments are essential for validation of laminarisation technology. They cannot completely be replaced by ground tests. Due to high cost they should be limited to a minimum. This may become possible by improvements in theory and ground test facilities, but some validation flight tests will remain essential.

18.8 Conclusion

Laminarisation is a promising technology for future Supersonic Commercial Transports. First investigations indicate significant improvements of aircraft efficiency.

Theoretical prediction of the transition laminar/turbulent requires substantially improved CFD-solutions of the undisturbed 3D-flow.

Stability analysis itself must be improved to describe coupling of different instabilities (CFI and TSI), but also to respect for non-adiabatic walls and HMI.

A quiet Ludwieg tube for Mach 1.6 to 2.4 and high Reynolds number (Re_L ca. 300 Mio) should be used in Europe to allow for high quality ground based check experiments.

A new design approach must be developed, based mainly on

- theoretical predictions
- and partial simulation by turbulent tests in classical supersonic wind tunnels
- but it must rely on selected checks in a quiet high Reynolds ground test facility
- and validation by carefully designed flight tests.

18.9 References

[376] **Mertens, J.**
Laminar Flow for Supersonic Transports in J. Szodruch (ed.): Proceedings of the First European Forum on Laminar Flow Technology.
Hamburg, 16.-18.3.1992, DGLR-Report 92-06

[377] **Boeing Commercial Airplane Company**
Application of Laminar Flow Control to Supersonic Transport Configurations.
NASA Contract Report 181917, July 1990

[378] **Stilla, J.**
Theoretische Transitionsvorhersage auf SCT-Tragflächen mit Überschallvorderkante. 9.
DGLR-Fach-Symposium der AG STAB, 4.-7. Oktober 1994, Erlangen, DGLR-Report

94-04, pp. 210-215

[379] **Mack, L.**
Boundary-Layer Linear Stability Theory AGARD Report No. 709 "Special Course on Stability and Transition of Laminar Flow", chap. 3, sub-chap. 9

[380] **Pfenninger, W.**
Flow Phenomena at the Leading Edge of Swept Wings.
AGARDograph 97 on "Recent Developments in Boundary Layer Research", Part IV, A. "Some Results from the X-21A Program, Part 1

[381] **Poll, D. I. A.**
Some Observations of the Transition Process on the Windward Face of a Long Yawed Cylinder. Journal of Fluid Mechanics 150 (1985), 329-356

[382] **Schrauf, G.**
Boundary Conditions for Compressible Linear Stability Theory.
Brite-Euram Project 1051/1064 "Investigation of Supersonic Flow Phenomena", Subtask 4.4: Transition Prediction, Final Report, June 1992

[383] **Arnal, D.**
Boundary-Layer Transition: Predictions Based on Linear Theory.
AGARD Report 793 "Special Course on Progress in Transition Modelling", chap. 2

[384] **Bippes, H.**
Experiments on the Influence of Surface Curvature upon the Development of Streamwise vortices in an unstable threedimensional boundary-layer flow. Z. Flugwiss. Weltraumforsch. **19** (1995) 129-138

[385] **Herbert, Th.**
Parabolized Stability Equations.
AGARD Report 793 "Special Course on Progress in Transition Modelling", chap. 4

[386] **Schrauf, G., Herbert, Th., Stuckert, G.**
Evaluation of Transition in Flight Tests Using Nonlinear Parabolized Stability Equation Analysis Journal of Aircraft, **33**, 3, May-June 1966, pp. 554-560

[387] **Herbert, Th., Schrauf, G.**
Crossflow-Dominated Transition in Flight Tests.
AIAA-Paper 96-0185, 34th Aerospace Sciences Meeting & Exhibit, January 15-18, 1996, Reno, NV, USA

[388] **Kleiser, L., Zang, T.**
Numerical Simulation of Transition in Wall-Bounded Shear Flows.
Ann. Rev. Fluid Mech. 1991. **23**: 495-537

[389] **Koppenwallner, G., Dorey, G.**
European Research and Testing Facilities Requested for Participation to SST/HST Projects. Proceedings of the European Symposium on Future Supersonic Hypersonic Transportation Systems, Strasbourg, 6-8 Nov. 1989, p. 306-316

[390] **Chen, F. J., Malik, M. R., Beckwith, L. E.**
Boundary-Layer Transition on a Cone and Flat Plate at Mach 3.5.
AIAA Journal **27**, 6 (June 1989), 687-693

[391] **Ludwieg, H., Hottner, H., Grauer-Carstensen, H.**
Der Rohrwindkanal der Aerodynamischen Versuchsanstalt Göttingen.
Jahrbuch der DGLR 1969, p. 52-58

[392] **Krogmann, P.**
An experimental study of boundary layer transition on a slender cone at Mach 5.
AGARD-CPP 224, Laminar-Turbulent-Transition, Lyngby, Denmark, 1977

[393] **Woan, C. J., Gingrich, P. B., George, M. W.**
CFD Validation of a Supersonic Laminar Flow Control Concept.
AIAA-paper AIAA-91-0188, January 1991

[394] **Aviation Week and Space Technology, December 2, 1991, p. 17**

[395] **Aviation Week and Space Technology, October 23, 1995, pp. 42-44**

[396] **The Air Letter**
Experiment smoothes supersonic wing flow. The Air Letter No. 13,598, Monday, 14
October, 1996, p. 7

CHAPTER 19

THE OBLIQUE FLYING WING TRANSPORT

A. Van der Velden

Synaps Inc., Atlanta, GA, USA

19.1 Introduction

In the last thirty-five years there have been numerous attempts to design an economical large long range supersonic transport aircraft. However, aircraft design teams around the world have not been successful at designing a 'Concorde' type large long range supersonic transport with realistic technology assumptions. It therefore seems only natural to look for other configurations that might fullfil this specification. The oblique flying wing configuration presented in this paper is unusual but we will show that it makes sense from both a technical and economical perspective.

Figure 115 shows the first oblique wing design. It was proposed by two Frenchmen, Edmond de Marcay and Emile Moonen around 1912 [402]. They saw the oblique position of the wing as a means to land the low speed aircraft of their time in the presence of crosswind without sideslipping the wing. In the decades to follow oblique wings were used to simulate sideslip in a windtunnel. It was not until 1943-1944 that engineers at Messerschmitt and Blohm & Voss combined the newly discovered phenomena of high speed drag reduction and wing sweep with the 'old' sideslipping wing. After WWII Richard Vogt of Blohm & Voss showed the oblique wing designs to R.T. Jones of NACA. Dr. Jones became convinced of the merits of this configuration and has actively pursued it until today. During the 1960's and 1970's he convinced many engineers at NASA, Boeing and Rockwell to study this exotic aircraft.

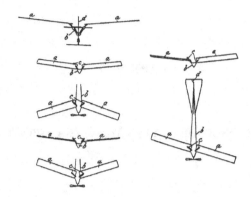

Figure 115   First Oblique Wing Design (Courtesy: Steve Ransom)

Figure 116 shows the AD-1 the first full scale oblique wing aircraft designed by Burt Rutan. On the 21 December 1979 the AD-1 this aircraft made its first flight. Though the oblique wing-body was a success from a controls and aerodynamics point of view, it had poor structural qualities. The increase in structural weight due to the routing of the loads through the center pivot was not offset by the high-speed drag reduction.

Figure 116   AD-1 by Rutan (1979)

In 1958, R.T. Jones and Lee of Handley Page [399] had proposed an even radical design that could overcome the high structural weight of the oblique wing body concept. Even though the design shown in Figure 117 was considered interesting, few have researched it since it was first proposed.

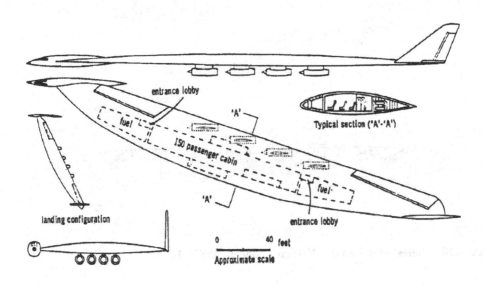

entrance lobby

fuel

150 passenger cabin

'A'

'A'

fuel

Typical section ('A'-'A')

landing configuration

entrance lobby

0   40 feet

Approximate scale

Figure 117   Lee's Slewed Wing Transport Proposal (1961)

The only research on this configuration - as known to the author - was done by Smith [401] in the U.K. and R.T. Jones in the U.S.A. As shown, the aircraft cannot be controlled due to its aft center of gravity position, even with todays artifical stabilization technology. Because the payload does not efficiently fit the available volume, the aircraft is too small for efficient super-sonic flight. As a consequence the configuration of Figure 117 was in many ways inferior to the Concorde, as Dr. Küchemann [398] correctly predicted at the time.

In 1987 Dr. Jones [397] and the author [403] proposed a new oblique flying wing design that takes advantage of the controls and oblique wing knowledge accumulated in the last four decades. In the last eight years we have learned far more about the design, but its basic lay-out has remained nearly unchanged. Figure 118 shows the latest version of the design. The present paper describes the state of the art of this oblique flying wing configuration and is largely based on the author's PhD thesis [405] at Stanford University.

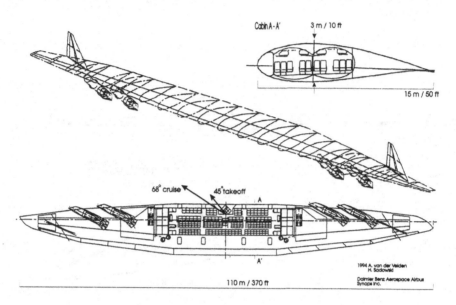

Figure 118   Jones' and Van der Velden's concept 1987 - 1994

19.2 Overview

To give the reader an idea of what an oblique flying wing is, we will start with a description the baseline design and layout.

Next we will focus on the main advantages of the configuration: Lower drag due to the variable sweep oblique wing and lower weight due to spanloading. It is the combination of both effects that make the oblique flying wing an economical long range supersonic transport.

A big concern for a new type of configuration is the question whether it can fly safely. In the third part we will therefore discuss the issues that relate to safe flight: controllability and stability. In particular we will discuss the results of the 1994 flight tests at NASA Ames.

In the 1970's the unacceptable environmental impact of supersonic transports, especially noise and ozone depletion were the cause of the cancellation of the USA SST program. So in the last part of this paper we will assess the oblique flying wings environmental performance.

19.3 Baseline Design and Layout

Figure 118 shows the present baseline Daimler Benz Aerospace Airbus Oblique Flying Wing layout designed for 250 passengers and a 5000nm range at a cruise Mach number of 1.6. The passengers are accommodated inside cylindrical hulls inside the constant chord center section. All oblique flying wings considered in this work have the same chord of about 15 m. At this chord the 19 % thick airfoil can hold two five-abreast hulls. The interior layout conforms to the A320 standard: its center aisle is 2.20 m high and the doors are 1.95 m high. Since passengers are accelerated sideways during takeoff, shoulder straps will be required. Other layouts in which passengers will not be accelerated sideways during takeoff and landing are optional. Entrance doors are placed in the nose of the aircraft. The emergency exits are found in the nose and trailing edge side of the passengers cabin, and can be reached by access ramps that lead to the top of the wing.

The pressurized hulls are laid along the spanwise direction and carry structural loads. The floor structure has a 50 cm crash zone. Figure 118 shows the cabin cross-section at the wing's zero incidence. The cabin is tilted with respect to the zero incidence so that the range of floor incidences never varies more than $3°$ from a level floor. The cargo is next to the passenger cabin, rather than under the floor, and offers space to containers up to 1.70m, so standard A340 LD3 containers can be fit.

Another deviation from the wide-body standard is the cockpit. Space is provided in the nose of the cabin to house two pilots. The pilot will have good visibility during approach and climb. His field of vision is $20°$ left, $90°$ right at takeoff and landing, similar to the co pilots view in FAR 25.777. The oblique wing is swept with the left tip forward so the pilots have an unobstructed right view. This is important with respect to the current air traffic right-of-way rules.

Figure 118 also shows the gear layout. The undercarriage has up to four main legs depending on maximum takeoff weight. The main landing gear is in the nose. A smaller gear is aft of the cabin. The best gear location is typically at 35% of the span for minimum structural weight. A constraint of 35 meters has been placed on the gear track so the aircraft can operate on runways of 50 m width. Figure 119 shows that it is necessary to have main gear steering to execute a turn. Takeoffs and landings are executed by rotation around the long aircraft axis. Apart from this unusual selection of the axis, the takeoff run is similar to that for a DC-3.

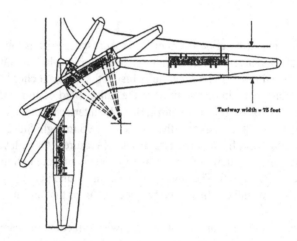

Figure 119   Runway Turn, Ref. [407]

Four fuel tanks are distributed over the span to reduce the structural loads during taxiing. To minimize trim drag, the center of gravity position can be moved by a fuel trim system as the case with all Airbus aircraft.

The nacelles are distributed along the span. The thrust-vectoring nacelle design shown in Figure 118 has the inlet almost parallel to the leading edge. It is comparable to the Olympus nacelle turned 90° around its long axis. At cruise the airflow is not turned, but at takeoff the inlet sucks in air from the right - just as Concorde at a high angle of attack sucks in the air from below the aircraft -. Inlet bypass doors prevent the air from separating in the nozzle. The nozzle then vectors the thrust to propel the aircraft at the takeoff sweep of 45°.

In view of the limitations of the artificial stability and control system the nacelles are placed as far forward as possible; synergistic, cabin noise and aerodynamic considerations dictate their placement outside the passenger cabin. Engine core or fan bursts do not cause damage to the pressure hull, therefore greatly reducing the critical risk of sudden decompression above 13000 m. To increase engine-out yaw control and to minimize the wave drag and wing stress, the engines were podded in four nacelles. The engines are of conventional turbofan design with a low bypass ratio. Such an engine placement would not lead to any significant additional drag (3 counts) for Mach numbers between 1.4 and 1.8.

This baseline configuration avoids one of the classical objections against the flying wing, namely that it does not have stretch potential, is not true for our baseline configuration. We can simply add center cabin sections of the wing's maximum thickness. It can be easily shown that this will even increase the L/D of the configuration. Although the OFW is very long in comparison to other aircraft, Figure 120 shows that it can fit a realistic airport slot. The current designs fit a Very Large Transport slot sized at 80 m x 80 m.

Figure 120   The Oblique Flying Wing at Munich II

Figure 121 shows a typical flight. It starts with the wing swept 45° and the engines streamwise to the flow. At the end of the takeoff run the cabin is rotated 6° around the long axis bringing the cabin floor up to 3°, and the aircraft takes off. The aircraft is than swept to 52° to decrease the gust loads and climbs to 13000 m. At this point the aircraft accelerates through Mach 1.0 and starts an accelerated climb to 15000m, sweeping the wing to 68° at Mach 1.6

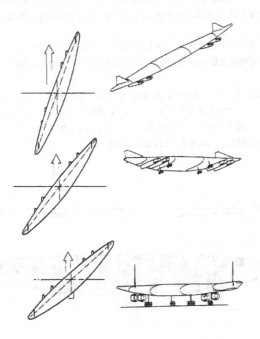

Figure 121   Oblique Flying Wing Views

19.4 Aerodynamics

Because the oblique flying wing can adjust its sweep angle for each Mach number, it achieves higher lift-to-drag ratios than conventional configurations up to Mach 2.0. A conceptually good way of looking at this is presented by:

$$Drag = Friction + \frac{L^2}{q\pi b^2} + \frac{(M^2-1)L^2}{q2\pi X_1^2} + \frac{128qV^2}{\pi X_2^4} \tag{108}$$

The first term in this expression, Friction, can be assumed constant. The second term is the induced drag. L represents the lift, q the dynamic pressure and b the aircraft span. The larger the span the less the induced drag for a given lift. The third term is the wave drag due to lift. M is the Mach number and X_1 represents the (weighted) average characteristic length of the aircraft's pressure signal. For low supersonic Mach numbers the lengths X_1 and X_2 are close to the projected length of the configuration in the direction of flight. V in the third term represents the volume of the configuration. So the reduce the wave drag it is necessary to have a very long aircraft.

Since we need both a great span and a great length to minimize the drag of a configuration it makes sense to consider a variable diagonal distributon of volume and lift. Because the sweep is variable such a configuration can minimize drag for any Mach number.

Thick supercritical sections are a key technology for the oblique flying wing. Sections as thick as 19 % will enhance the utilization of the wing volume reduce of structural weight.

Figure 122 shows the example resulting from the author's direct wing design method as described in chapter 16, see the OFW case study there (Figure 93). In view of the limitations of the artificial stability and control system the wing was trimmed around 32% of the chord. The wing is naturally stable in front of this location.

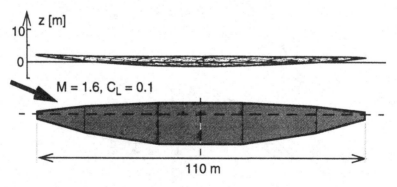

Figure 122   Optimized aerodynamic shape for case study (Figure 118).

The global design (wing area, thickness, aspect ratio etc.) was optimized for minimum total operating cost [406] . The detailed design originated out of the optimization of the shape for minimum drag at cruise without moments around the center of gravity. Additional constraints were imposed to fit the payload according to the standard previously discussed. The design clearly shows the wing parabolic dihedral (bend) proposed by R.T. Jones. A bent oblique wing creates a linear twist distribution which compensates for the loading up of the aft wing due to induced effects.

Figure 123 - Figure 127 show the aerodynamic forces and moments during flight for the full scale transport using this optimized shape. Notice that the aircraft flies only in very narrow corridors of angles of attack. The lift-to-drag ratios are in excess of 25 at subsonic speeds and in excess of 11 at supersonic speeds. The pitching and rolling moments cannot be decoupled: if one analyzes the moments involved, it turns out that the neutral point moves along the center of gravity line. This is caused by the upwash induced from the forward wing on the aft wing. For symmetric swept aircraft this phenomenon only causes increased pitching moments, but for oblique wings this causes rolling moments.

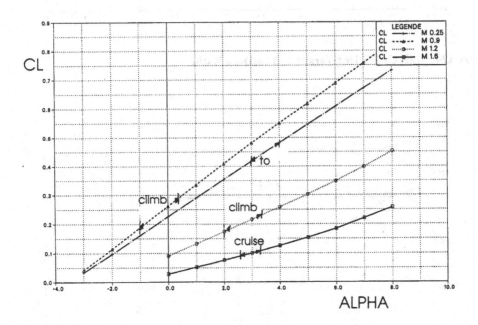

Figure 123   Angle of Attack Versus Lift Coefficient

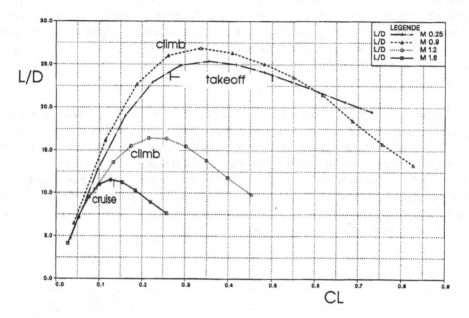

Figure 124   Lift Coefficient Versus Lift-to-Drag Ratio

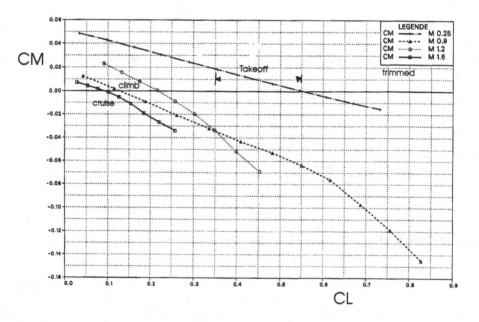

Figure 125   Lift Coefficient Versus Pitching Moment Coefficient

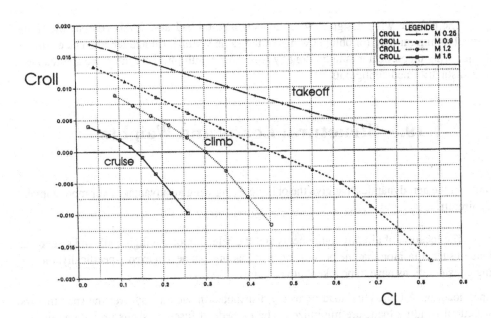

Figure 126   Lift Coefficient Versus Rolling Moment Coefficient

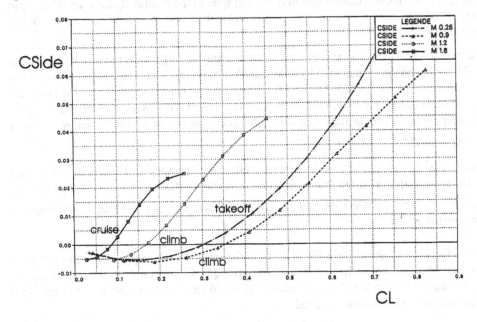

Figure 127   Lift Coefficient Versus Side Force Coefficient

So, in a sense, the aircraft is close to neutrally stable about the long axis. Under normal flight conditions there is no side force on the aircraft; at higher angles of attack, the leading edge suction produces a side force to the right.

The oblique wing designed with this new method had drag levels comparable to the theoretical minimum drag for oblique wings of the same volume, lift and length. The estimates of oblique wing theoretical minimum drag by Jones and Smith [401] are therefore applicable to more realistic trimmed configurations.

19.5 Structural Loads and Weights

It is not just the aerodynamics that give the oblique flying wing an advantage over more conventional aircraft.

The location of aircraft components and the selection of the flight path and sweep can be done in such a fashion that the benefits of span loading can be achieved. Specifically, the following considerations have to be taken into account:

- Span loading. All mass and load items are distributed in such a way over the span that the structural bending loads are minimized. The embedded fuselage shells provide additional bending stiffness in the middle of the configuration. Figure 128 shows the smeared equivalent skin box thickness distribution over the half span of the oblique flying wing. Thus, span loading allows for an almost constant required skin thickness of 7mm allowing for efficient manufacturing.

- Runway loads. The loads experienced during taxi and landing can be reduced by a two-legged main landing gear supported by a two auxillary aft gears. Such a layout also provides stable maneuvering during taxi. To allow the oblique flying wings to grow to the span required for efficient supersonic flight (120m) the gear legs have to be at least 40 to 50 meters apart. In order not to exceed the runway width such a gear span can only be achieved by sweeping the wing to 45°.

- Gust loads. To reduce the gusts loads during climb the operation of the aircraft is restricted to lower equivalent airspeed than typical transonic transports. In addition the wing will have to be swept to 50° to reduce the high lift gradients that cause high gust loads and to improve ride quality further.

Operation of the oblique flying wing in this fashion will increase the runway length and increase drag but not to the extent that this outweighs the structural advantages created by them. The advantages created by load minimization and span loading account for at least half the total economic benefits that are cited.

On top of the benefits of spanloading the oblique flying wing will also profit more from the application of composites materials then current subsonic aircraft because of the higher productivity per pound of structural weight for a supersonic aircraft. However, in the case of the OFW their use is not required. Industry experience shows that intermediate carbonfibers with BMI resins achieve strain levels in excess of 0.5 % can result weight savings over conventional primary structures in excess of 25 %. The airframe life was specified to be over 75,000 hours with 50,000 supersonic flying hours and 25,000 pressure cycles. A minimum skin thickness of 2 mm was specified to minimize foreign object damage.

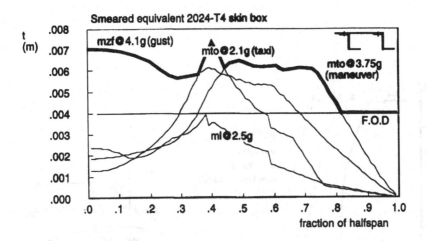

Figure 128   The Smeared Equivalent Wing Box Thickness

19.6 Stability and Control

19.6.1 Concept

A control system consisting of a narrow trailing edge flap and at least one vertical fin can provide sufficient control authority to trim the aircraft at the 32 % chord design center of gravity in both supersonic cruise and takeoff and landing. Only a very small (typically 3 %) aerodynamic center shift was observed going from subsonic to supersonic speeds.

Stability and control in pitch and roll is provided by a 10% multi segmented trailing edge flap, segmenting this flap increases the system reliability and enables roll control; each flap segment can be independently controlled by the on board flight computers. This flap system will put the neutral point as far back as 37% of the mean aerodynamic chord at OEW (operating empty weight), and smooth out any gust peaks.

The artificial stability and control system that controls this flap may use a standard feedback controller. This controller relates the aircraft attitude and attitude rate of change to an optimum flap deflection.

Figure 129 shows the predicted rearward stability limits when such a feedback system is in place. (The dynamic model used in this work was only quasi-3d and accounted for aerodynamic lag.) The dynamic stability limit is set by a 20.13 m/s (66ft/s) gust at minimum control speed. As can be seen, the system is more sensitive to step gusts than to an FAR25 (1-cosine) gust. The rearmost center of gravity position is located in front of the step gust's neutral point.

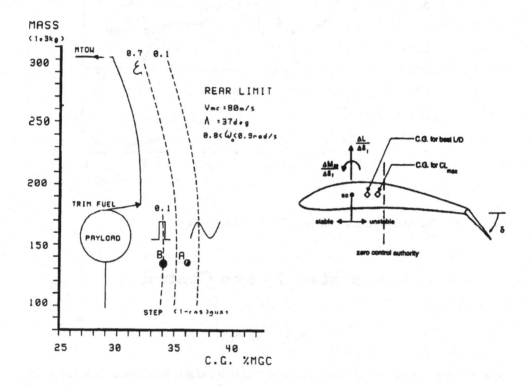

Figure 129   Rearward Stability Limits

The stability limit moves forward with increased configuration weight because the flap deflection required to balance the aircraft will cause the flap to stall for limit gusts in low speed flight. Typical aircraft responses for center of gravity positions close to the neutral point are depicted in Figure 130 (A and B refer to conditions in Figure 129).

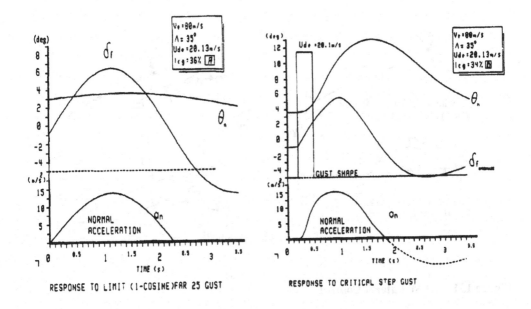

Figure 130   Step and Cosine Gust Response

The configuration has at least one "all-flying" vertical fin. The combined size of the vertical tailplanes is set by the two engine out condition at takeoff. As shown in Figure 131, both vertical tailplanes are deflected to oppose the yawing moment caused by the inoperative engine. Crosswind landings can be performed without a bank angle. The vertical tailplanes and engines are turned in the direction of the sideslip to eliminate the side force created by the sideslip angle. For sideways maneuvering the inboard front and rear vertical tail should be loaded equally to balance the configuration around the z-axis. The present configuration can manage with C_Y's up to 5% of C_L by vertical tail deflection alone. Even more side force can be generated by banking the wing.

Figure 131 also shows the required control deflections to counter a vertical gust. An upward vertical gust increases the angle of attack which in turn causes increased leading edge suction and side force. The side force can be compensated by symmetric vertical tail deflection.

19.6.2 Proof of Concept Demonstrators

In the summer of 1987, Steve Morris flew an unpowered model of an unswept flying wing with such an artificial stability augmentation system. The model was dynamically stabile with a center of gravity at 32% of the mean aerodynamic chord. In 1990 Dr. Morris built and flew a 10 ft span powered oblique flying wing that was naturally stable. This aircraft was flown at sweep angles up to 60 degrees.

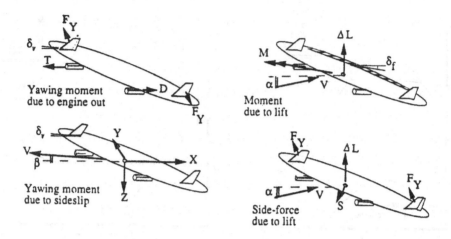

Figure 131   OFW Lateral Control

After these initial successes, NASA Ames awarded a $ 250,000 two-year grant to build a 20 ft, two engined oblique wing demonstrator. Even though its wingspan almost matches that of a Cessna 152, the model OFW is still relatively small compared to the 400 ft full scale aircraft. This model is shown in Figure 132; the layout of the model is shown in Figure 133. Its flight stability is determined by the quantities $\mu = \frac{mass}{\rho Sl}$ and $K^2 = \frac{inertia}{mass \cdot l^2}$ as well as the aerodynamic force and moment derivatives.

Figure 132   Twenty Foot Oblique Flying Wing Model (Courtesy S. Morris)

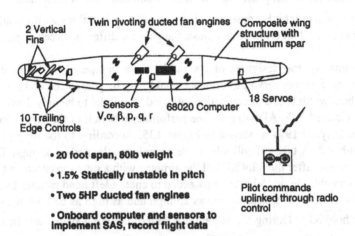

Figure 133   Layout of the Twenty Foot Oblique Flying Wing Model (Courtesy S. Morris)

Both the full-size aircraft and the model have the same values of these parameters at takeoff. The aerodynamic derivatives are worse for the model because of Reynolds number effects. Figure 134 shows the calculated OFW off-design performance with stability augmentation in place.

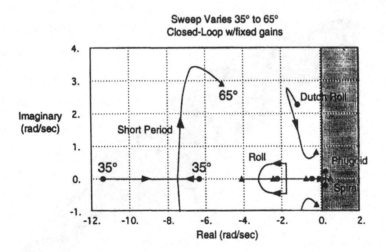

Figure 134   Off-Design Performance with Stability Augmentation (Courtesy S. Morris)

The model, as designed, relied on simplified control laws that used only 6 fixed gains and made the assumption that pitch and yaw were decoupled. The current model aircraft is stable up to 50° at which point it becomes spirally unstable. The spiral instability could be controlled by a different control algorithm that adjusts the gain for different sweep angles.

Unfortunately, the designers of the model were hampered by lower than expected servo performance. Reliable servo performance could only be guaranteed at 3 Hz. The severely reduced servo bandwidth (10Hz specification) allowed the model to be only flown 2 % unstable rather than the planned 7 %. Allowing for the performance deficit of the servos the model flew as predicted on May 10 1994 as shown in Figure 135. According to the report [400] : " The flight began with a 23 s take-off roll where the aircraft accelerated to 45 mph TAS and then lifted from the ground after the pilot rotated the airplane with a pitch-up command... The model then climbed to an altitude of 150 ft and proceeded to enter a left hand pattern. During this pattern the model was flown at speeds as low as 25 mph and as high as 65 mph with wing sweep commanded to hold 35° . During the second left pattern the wing sweep was increased to 50° for a few seconds and then returned to 35° (as shown in Figure 136). At the end of the second pattern ... the landing pattern was easily executed and the model landed safely on the runway centerline direcly in front of the the video cameras and the flight data was retrieved. The pilot commented that the airplane flew more easily than the 10 ft. model which had no on-board computer.

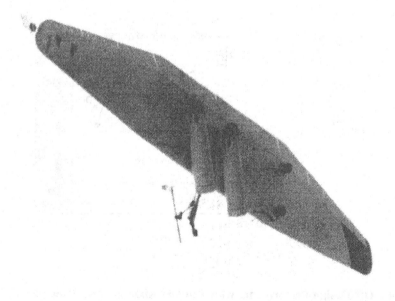

Figure 135   The Twenty Foot Oblique Flying Wing Model in Flight

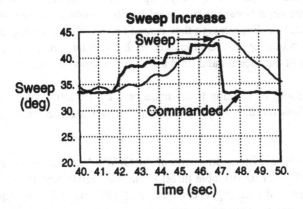

Figure 136   Response to Command During Turning Flight

19.7 Environmental Impact

19.7.1 Noise

Strictly enforced noise regulations make the introduction of a long range supersonic delta wing type configuration almost impossible. The reason for this can be 'easily' understood.

Since the noise produced depends on the aircraft thrust, and since the thrust has to be balanced by the drag, higher drag means more thrust and thus more noise at a given airspeed. The large supersonic delta wing aircraft designs have take-off weights that are comparable to those of the 747 and strive for the same take off field lengths. Unfortunately their wing spans are typically only 60 % of the 747's in order to make the wing longer for the same wing area. This reduced span more than doubles drag at the same flight speed.

If we look at the equation for induced drag in the aerodynmics section we see that the designer can also increase the takeoff speed to lower the induced drag. Higher takeoff speeds at the same thrust increase the takeoff field length and reduce the climbout distance to the noise flyover point. The designer may thus satisfy the sideline noise regulations, but not the flyover regulations or the field length constraints. If the designer decides to increase the thrust, the will reach rotation sooner and will climb out higher. Now he satisfies the flyover regulations, but not the sideline regulations.

So for an aircraft of the same weight and less span we need an engine that makes less noise at a given thrust. This can only be achieved by increasing the massflow through the engine. Therefore a long range supersonic delta wing type configuration will have to be fitted with engines that have a much greater crosssection that B747 engines at takeoff. These extremely wide engines also have to be very long because of the supersonic compression process. These very large engines cannot be spaced to close together because then they will start sucking in each others air. Even though they cannot be spaced to close together, they will have to be fitted on a much smaller span. On top of all of this such large massflow engines can double the weight and drag of the nacelles and limit the takeoff rotation angle.

The author sees no solution to the noise problem for large delta wing supersonic aircraft at this time. However, the oblique flying wing is not caught in this catch 22 of noise and performance.

The high subsonic lift-to-drag ratio of the OFW allows the bypass engine to be throttled back to about 50 % of its available thrust while still maintaining the required airfield performance. Such a throttled back engine can meet the FAR36 stage 3 noise requirements without significant noise abatement measures. Future noise regulations will be met with similar penalties as those experienced by competing subsonic transports.

19.7.2 Sonic Boom

Sonic boom cannot be avoided. However current large supersonic delta wing type designs produce sonic booms in excess of 2.5 psf, much more than Concorde. On top of that the subsonic specific range of these aircraft is less than the supersonic cruise specific range. Flying subsonic over land will therefore result in a significant range penalty.

The sonic boom of an oblique flying wing is much less than Concorde's. It was determined using the Whitham F-function method and the TranAir full potential code. The OFW was modeled by a slewed elliptic lift and Sears-Haack area distribution, a panel method and a high definition surface geometry. All representations gave similar boom signatures, as follows:

- Bow shock over pressures between 50 to 80 N/m^2 depending on aircraft size and mission segment.
- The aft-shock is canceled due to favorable volume-lift interference.
- The lateral distribution of the sonic boom signature is distinctively asymmetric.

The sonic boom is still too loud for unrestricted flight. The best way to improve the sonic boom is to eliminate it completely by flying below Mach 1.2. Unlike other proposed HSCTs the variable geometry oblique flying wing can cruise at these speeds with very high fuel efficiency. In terms of the overall operational loudness, the larger OFW's provide an improvement over the smaller OFW's because of the reduction of the number of sonic booms for the same production of seat kilometers.

19.7.3 Ozone

The effect on ozone depletion by supersonic transports is still a hotly debated issue. But most scientists accept that ozone depletion by nitrous oxides is caused by the following mechanisms in order of importance:

- Cruise altitude. For a given rate of NO_x injection into the atmosphere the rate of ozone depletion will go up nearly linearly with altitude from altitudes between 15 and 20 km.

- Combustor entry conditions. The higher the combustor entry temperature and pressure the higher the formation of NO_x.

- Fuel Efficiency. The higher the fuel consumption the higher the formation of NO_x.

Though the lower wing loading contributes to a higher cruise altitude, this effect is almost completely offset by the reduction in parasite drag. The reduced parasite drag will lead to a lower cruise lift coefficient and therefore a lower cruise altitude. Because the effect of fuel efficiency is less important than lowering the combustor entry conditions, the power plant efficiency will be penalized by very strict ozone depletion standards. After all elements were taken into account we did not find a significant difference in the ozone depletion of conventional wing-body aircraft and the oblique flying wing when the same cruise Mach number was considered. The effect OFW's improved fuel efficiency was canceled by its somewhat increased flight altitude. According to the Chang model, the impact of a fleet of Mach 1.6 OFW's replacing the current fleet of B747's on the ozone layer will be less than the 2.5 % reduction of the ozone column proposed by NASA. This is about one fourth of the impact of Mach 2.4 conventional transport and ten times the impact of the current subsonic fleet.

19.8 Economy

The aerodynamic and structural benefits of the oblique flying wing translate directly into better aircraft operating economy.

The oblique flying wing was directly compared with conventional supersonic and subsonic transports based on operating economy for a range of missions with specified performance and environmental constraints. All aircraft were evaluated with the same analysis routines and to the same level of structural, aerodynamic and propulsion technology comparable of that achieved by the new generation of subsonic transports. In terms of direct operating costs, oblique flying wings with more than 400 passengers were superior to conventional wing-body configurations over the entire Mach 0.8 to Mach 2.0 operating range. The improvement was smaller than the uncertainty of the analysis at Mach 0.8, but more than a factor of two at supersonic speeds up to Mach 2.0. Unlike conventional delta wing transports it will be possible to design oblique flying wings for payloads up to 600 passengers and ranges up to 12000 km (6700nm) while satisfying current economic, performance and environmental requirements.

Table 6 compares a 400 passenger OFW and a 747-400.

	OFW	B 747-400
Cruise speed	M 1.6	M 0.85
Range	9000 km	10300 km
Geometry:		
Wing Area m^2	1316	511
cabin l x w (m)	39.2 x 7.5	57 x 6.1
total l x w (m)	120 x 14.5	69 x 60
Weights:		
OE (kg)	127000	180000
MTO (kg)	30700	385000
SLS Thrust (kN)	4 x 230	4 x 250
Price (1994)	285 M$	140 M$
Production (seat.km.year)	3.0×10^9	1.8×10^9
Est. Total Operating costs	9 $ct / pax.nm	9 $ct / pax.nm

Table 6 Comparision of an OFW with a 747-400

Figure 137 to Figure 139 compare the total operating cost of optimized oblique flying wings with other optimized aircraft configurations.

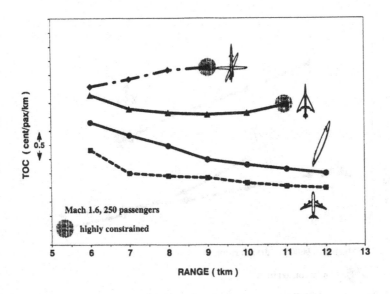

Figure 137   Comparison of TOC's as a Function of Range

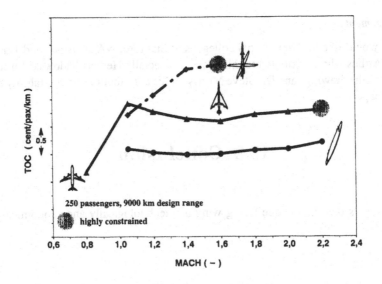

Figure 138   Comparison of TOC's as a Function of Speed

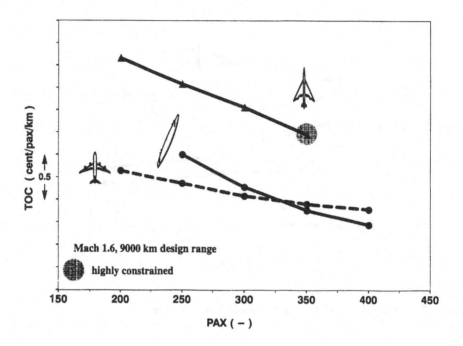

Figure 139   Comparison of TOC's as a Function of Size

Acknowledgements

The author would like to thank all his collegues at Stanford, NASA Ames and Daimler-Benz Aerospace Airbus who contributed to this project, especially Herbert Sadowski for his work on the layout and the drawings and Dr. Steve Morris for his contributions to the stability and control part of this paper.

19.9 Conclusion

This paper shows that the oblique flying wing is a technologically and economically feasible transport.

19.10 References

[397] Jones, R. T.
The Supersonic Flying Wing, Aerospace America, November 1986

[398] Küchemann, D.
The Aerodynamic Design of Aircraft, Pergamon Press, 1979

[399] Lee, G. H.
Slewed Wing Supersonics, *The Aeroplane*, March 1961

[400] Morris, S. J., Tigner, B.
Flight Tests of an Oblique Flying Wing Small-Scale Demonstrator, Stanford University, unpublished, September 1994

[401] Smith, J. H. B.
Lift/Drag Ratios of Optimised Slewed Elliptic Wings at Supersonic Speeds, *The Aeronautical Quarterly*, August 1961

[402] Ursinus, O.
Flugzeug mit schwingbaren Tragflächen, Flugsport no. 21, 15 Oktober 1913 (information courtesy of Steve Ransom)

[403] Van der Velden, A. J. M.
The Conceptual Design of a Mach 2.0 Oblique Flying Wing Supersonic Transport, NASA CR 1777529, May 1989 (edited version of the December 1987 report)

[404] Van der Velden, A. J. M., Kroo, I.
The Sonic Boom of an Oblique Flying Wing, *Journal of Aircraft*, Jan.-Feb. 1994.

[405] Van der Velden, A. J. M.
Aerodynamic Design and Synthesis of the Oblique Flying Wing Supersonic Transport, PhD-thesis Stanford University, Dept. Aero Astro SUDAAR 621, Univ. Microfilms no. DA9234183, June 1992

[406] Van der Velden, A. J. M., Von Reith, D.,
Multi-Disciplinary SCT Design at Deutsche Aerospace Airbus , 7th European Aerospace Conference EAC '94. 25-27 October 1994 Toulouse.

[407] Waters, M., Ardema, M., Kroo, I.
Structural and Aerodynamic Considerations for an Oblique All-Wing Aircraft, Aircraft Design, Systems and Operations Conference AIAA 92-4420, Hilton Head August 1992.

CHAPTER 20

OBLIQUE FLYING WING STUDIES

A.R. Seebass

University of Colorado, Boulder, CO, USA

20.1 Introduction

In 1968, as mentioned in Chapter 1, there were 97 thousand aircraft arrivals and departures at Kennedy International Airport providing air travel for nearly 8 million passengers. By 1993, 15 million international passengers used Kennedy, but this required only 92 thousand arrivals and departures. The introduction of the Boeing 747, DC 10 and other large aircraft, starting in 1970, made this possible. Kennedy, and many other airports, are near their capacity. This, in part, is the motivation for large transport studies. One aircraft is naturally suited to being large and to flying supersonic: the Oblique Flying Wing (OFW). This aircraft represents a radical departure from past configurations.

The Oblique Flying Wing (OFW) as a moderate size SST is considered in detail in Chapter 19. As a sequel to that chapter, we review some important aerodynamic results for the OFW. Next we review other studies of the OFW. We then provide new aerodynamic results of our own that derive largely from the application of the theoretical tools developed in Chapter 7 and the geometric tools of Chapter 9. We conclude by delineating the advantages and disadvantages of an Oblique Flying Wing.

20.2 Linear Theory

The supersonic area rule tells us that the wave drag of an aircraft in a steady supersonic flow is the average wave drag of a series of equivalent bodies of revolution. These bodies of revolution are defined by the cuts through the aircraft made by the tangents to the fore Mach cone from a distant point aft of the aircraft at an azimuthal angle θ. This average is over all azimuthal angles, as depicted in Figure 140 for an OFW. For each azimuthal angle the cross-sectional area of the equivalent body of revolution is given by the sum of two quantities: the cross-sectional area created by the oblique section from the tangent to the fore Mach cone's intersection with the aircraft, projected onto a plane normal to the freestream; and a term proportional to the component of force on the contour of this oblique cut, lying in the θ = constant plane, and normal to the freestream.

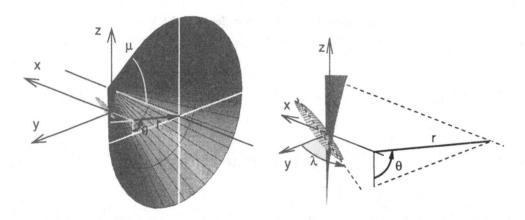

Figure 140   **Fore-Mach cone (left) for the linearized solution, intersecting the OFW: detail right.**

For minimum wave drag due to lift, each of the equivalent bodies of revolution due to lift must be a Kármán ogive [408]. Likewise, for minimum wave drag due to volume, or due to thickness, each of the equivalent bodies of revolution due to volume, or due to caliber, must be a Sears-Haack, or a Sears, body [409], [410]. Because the wing also serves as the fuselage, the OFW should have a small wetted area as well. These conditions can be met by an elliptical wing flying obliquely, so that its normal Mach number is subsonic, as Jones and Smith pointed out long ago [411], [412]. An ellipse has the important characteristic that its chord length distribution is elliptical regardless of the angle of the chord to the center line. For minimum wave drag due to volume, the thickness distribution must be parabolic; for minimum wave drag due to thickness, the thickness distribution is more complex.

At supersonic speeds the wing must be swept to a sufficient angle that it functions efficiently as a high Mach number subsonic wing in the cross-flow normal to the wing. This then requires a wing derived from supercritical airfoils, or a full supercritical wing design. Since the wing must house the passengers, it must be a relatively thick wing in order not to be so large as to be impractical.

The drag of an oblique elliptic wing can be expressed using our theoretical understanding of drag at supersonic speeds. The individual components of drag are the skin friction drag, the induced drag, the wave drag due to lift, and the wave drag due to volume. The lower bound for this drag, for a given lift and volume, is given by

$$D = qS_fC_f + \frac{L^2}{\pi qs^2} + \frac{\beta^2 L^2}{\pi q l_l^2} + \frac{128qV^2}{\pi l_v^4} . \tag{109}$$

Here q is the dynamic pressure of the free stream, S_f the reference area for skin friction, and C_f the skin friction coefficient. The first term represents the wing's skin friction drag, which we may reasonably approximate by the turbulent drag on a rectangular flat plate of the same area and streamwise chord.

The second term is the induced drag for an elliptically loaded wing, where L is the lift. We recognize this expression as the induced drag of the wing in the flow normal to it, that is, as $L^2/(\pi q_n b^2)$, where q_n is the dynamic pressure of the normal flow and b is the unswept wing's span. We then interpret the product $q_n b^2$ as $q\cos^2\lambda\, b^2$ or qs^2, where λ is the sweep angle, and s is the span normal to the free stream.

The third and fourth terms are the minimum wave drag due to lift and volume, where V is the wing's volume and $\beta^2 = M^2 - 1$. The two lengths, l_l and l_v, are the averages over all azimuthal angles of the individual lengths of the equivalent bodies of revolution, appropriately adjusted for the variation of the component of the force lying in the θ = constant plane.

If the loading in one oblique plane, say spanwise, is elliptical, and the wing has an elliptic planform, the loading will be elliptical in all azimuthal planes. To obtain an elliptic loading will require one or more of: wing bending; twist variation; camber variation.

If such an elliptic wing has a parabolic thickness distribution, the equivalent body due to volume in all azimuthal planes is that of a Sears-Haack body. The wing's thickness is set by passenger height. If we minimize the wave drag due to the wing's thickness, that is due to the caliber of the equivalent body, then for the same caliber body as the Sears-Haack body, the drag is given by Eq. (109) with V reduced by $\sqrt{(8/9)}$.

As noted above, the lengths in the last two terms are the average over all azimuthal angles of the *effective* lengths for lift and volume for each azimuthal angle, as determined by the supersonic area rule. Thus l_l is the actual length for that azimuthal plane divided by $\cos\theta$. To calculate these lengths we must determine the angle at which the tangent to the Mach cone cuts the plane of the wing.

For simplicity we assume that the wing lies in a horizontal plane. We recognize, but ignore, the fact that the wing must incline its lift vector slightly or be otherwise trimmed to offset the leading edge suction which occurs on only one side of the wing. This results in wing plane inclination of about one degree.

If we write down the expression for the fore Mach cone depicted in Figure 140, and consider its apex to be at a large radial (and thereby axial) location, this equation becomes that for its tangent plane. This plane intersects the horizontal plane, and thereby the wing, in a line that makes an angle with the y-axis, φ, given by

$$\tan\varphi = \pm\beta\sin\theta. \tag{110}$$

Now that we know the angle cut by the tangent to the Mach cone, we also know the length of the equivalent body of revolution for that plane is

$$l(\theta) = b(\sin\lambda - \beta\cos\lambda\sin\theta). \tag{111}$$

The loading on the cut made through the wing is composed of two parts: that due to the lift, and that from the leading edge suction; we ignore this latter force. With these approximations we may determine the two lengths l_l and l_v [413]:

$$\frac{1}{l_l^2} = \frac{1}{2\pi}\int_0^{2\pi}\frac{(\cos\theta)^2}{l^2(\theta)}d\theta = \frac{1}{m^2b^2(\sin\lambda)^2}\left(\frac{1}{\sqrt{1-m^2}}-1\right); \tag{112}$$

$$\frac{1}{l_v^4} = \frac{1}{2\pi}\int_0^{2\pi}\frac{d\theta}{l^4(\theta)} = \frac{2+3m^2}{2b^4(\sin\lambda)^4(1-m^2)^{7/2}}. \tag{113}$$

Here $m = \beta\cot\lambda$; for a wing with subsonic leading edge, $m < 1$; for large sweep, that is small m, Eq. (112) gives $l_l^{-2} = l(0)^{-2}/2$. Identifying $l(0)$ as the wing length, we see that eq. (109) is equivalent to Eqs. (50) and (108) in the previous chapters 4 and 19.

The minimum drag arising from the lift of a wing in supersonic flight, i.e., the sum of the induced drag and wave drag due to lift, was first given by Jones [411]. It can also be determined by applying Kogan's theorem [414]. And this theorem can also be used to show that an oblique, elliptically loaded wing has the minimum inviscid drag for a given lift [415].

We may use Eqs. (109) and (112) to determine the inviscid drag of an oblique lifting line. This gives

$$D = \frac{L^2}{\pi q s^2\sqrt{1-m^2}}. \tag{114}$$

This is the result first derived by Jones, using the principle of combined flows [411],

[416]. The linear result for an arbitrary elliptic wing is more complex. We give the result here for an oblique wing of large aspect ratio [413]:

$$\frac{C_D}{C_L^2} = \frac{(\cos\lambda)^2(1+m^2)}{\pi A(m^2+G)\sqrt{1-m^2-G}}$$ (115)

where m is related to the sweep angle and aspect ratio by

$$m = \beta\cot\lambda\left[1-32\left(\frac{(\cos\lambda)^2}{\pi\beta A\cos 2\lambda}\right)^2\right],$$

and G, which vanishes for a lifting line, is

$$G = \frac{1}{1-m^2}\left[\frac{4(\cos\lambda)^2(1+m^2)}{\pi\beta A}\right]^2.$$

Here A is the swept wing's aspect ratio, that is, s^2 divided by the wing area.

An oblique elliptic wing simultaneously provides large span and large lifting length. The reduction in the wave drag of an oblique wing of finite span also comes from being able to provide the optimum distribution of lift in all oblique planes.

Figure 141 and Figure 142, taken from Jones [417], illustrate the aerodynamic advantages of the oblique wing. The first compares the minimum inviscid drag due to lift of an oblique elliptic wing with that of a delta wing of the same span and streamwise length. The second figure compares the wave drag due to volume of an oblique elliptic wing and a swept wing having the same aspect ratio and the same thickness.

$C_D/C_L^2 = .15$ $C_D/C_L^2 = .30$

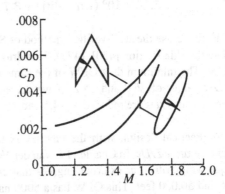

Figure 141   (left) Drag due to lift; oblique elliptic wing and delta wing, $M = \sqrt{2}$ (Ref [417]).
Figure 142   (right) Wave drag due to volume of oblique elliptic and swept wings as a function of Mach number (Ref. [417]).

To fix our ideas on the relative size of the various contributions to the total drag, and the possible maximum L/D let us consider a concrete example.

20.3 The OFW as the NLA

Studies by McDonnell Douglas Aerospace West provide guidance on how many passengers a large OFW might carry, how much it might weigh, and at what speeds and altitudes it might fly for the Mach number range 1.3-1.6 [418], [419]. We choose a freestream Mach number of $\sqrt{2}$ for simplicity, and a sweep angle of 60 degrees for ease of control and aeroelastic stability. Higher speeds are possible with more sweep, but the wing's control becomes increasingly difficult, with 60 degrees being judged acceptable in previous studies.

These parametric studies guide us to conclude that a high aspect ratio, 800 passenger aircraft will have a wing with a maximum chord of about 55 feet and a span of about 550 feet. This 10 -1 maximum chord to span ratio provides an aspect ratio of 12.7, a wing area of 23,758 square feet and a wing volume of 127,815 cubic feet. This OFW should have a trans-Pacific nautical mile range.

A conservative guess as to its weight and volume provides an OFW transport that will enter cruise at about 1.9 million pounds and leave cruise at a weight of 1.3 million pounds. For nominal conditions we take the weight to be 1.6 million pounds and the cruise altitude to be 43,500 feet.

Using these results and the nominal conditions, we calculate the turbulent skin friction drag on a flat plate, and more directly, the other terms to conclude that the drag in pounds is:

$$D = 4.37 \times 10^4 \text{ (skin friction)} + 2.31 \times 10^4 \text{ (induced)}$$

$$+ 5.19 \times 10^3 \text{ (wave-lift)} + 3.74 \times 10^4 \text{ (wave-volume)},$$

Here we use the well-verified method of Sommer and Short, assuming adiabatic flow, to calculate the skin friction [420], [421]. This drag gives an inviscid L/D of 24.4 and a viscous L/D of 14.6. The minimum drag OFW of the same thickness (but less volume) has inviscid and viscous L/D values of 26.0 and 15.2. If we correct this lifting line result using Eq. (92), the respective viscous L/Ds are reduced to 14.5 and 15.1.

A practical design, with the engines in the wing and only one vertical fin, should nearly achieve these L/Ds. In Chapter 19, Van der Velden describes in some detail a 250 passenger oblique wing with four external engines that cruises with its wing swept to 68 degrees at Mach 1.6 and 50,000 feet. This OFW has a 5000 nautical mile range. Its 19% thick wing has a maximum chord of about 50 feet and a span of about 370 feet. Its maximum cruise L/D is about eleven. Because the component of the Mach number normal to the wing is only 0.6, a 19% thick wing should be possible.

20.4 Other Oblique Flying Wing Studies

The recent interest in OFWs derives from studies at Stanford by R.T. Jones, A. Van der Velden and his thesis advisor, I. Kroo. The early Stanford studies by Jones led to studies by the Systems Analysis Branch at NASA Ames, a Boeing in-house assessment of the concept, and contractual studies funded by NASA at Boeing, McDonnell Douglas, and Stanford. These culminated in wind tunnel tests of two candidate wings as well as small radio controlled models to evaluate low speed stability and control issues.

An excellent synopsis of the NASA supported work is provided by Galloway et al., who describe these studies and discuss the conclusions NASA drew from their own, and the contracted investigations [422]. Some of the discussion that follows derives directly from this report. A NASA artist's concept of an OFW is shown in Figure 143.

20.4.1 Stanford Studies

As his Ph.D. thesis at Stanford, Van der Velden undertook the development of a general evaluation tool for preliminary design of commercial supersonic aircraft, including the OFW [423]-[425]. This resulted in the preliminary design of an OFW that provided additional impetus and guidance to the NASA Systems Analysis Branch in their own studies of such aircraft.

Morris, in his Stanford Ph.D. thesis, examined the integrated aerodynamic/control system needed for a rigid oblique wing aircraft (OWA), that is, an oblique wing with fuselage [426]. He showed that by tilting the pivot axis, adverse coupling could be reduced and handling qualities improved. He subsequently extended these studies to the integrated aerodynamic and control system design of an oblique flying wing. The configuration matched that of a NASA design for a 400 passenger OFW [427]. Morris noted the utility of designing the control law for the principal axes. He then demonstrated this control law with two radio-controlled OFWs. The first was ten feet in span and flew successfully at up to 65 degrees sweep. The second was a twenty foot span OFW powered by two pivoting, 5 horsepower, ducted fan engines. Ten 25% chord trailing edge flaps were controlled by a Motorola 68020 CPU and a 68881 math coprocessor. The aircraft was designed to be able to fly at sweep angles from 35 to 65 degrees. Flight sensors were a 3-axis rate gyro, a 2-axis wind vane and an airspeed indicator.

Figure 143 **A NASA artist's concept of an OFW.**

In addition to the ten trailing edge control surfaces, two flying vertical fins, two throttles and four landing gear struts were driven by commercially available actuators. Actuator bandwidth limited the static stability margin to -1.8%. This aircraft flew in May 1994, successfully completing a four minute flight circling the field twice, changing its sweep from 35 to 50 degrees and back. It circled the field by turning toward the trailing tip, which resulted in sweep angles as low as 20 degrees because of the high damping in yaw. Morris, his model aircraft, and some of his results are depicted in some Figures of Chapter 19.

20.4.2 Boeing Studies

After their own in-house study of an OFW transport, the Boeing Company, under contract from NASA, developed an OFW design that would fit current airport designs and meet current FAA requirements. To satisfy the FAA requirement that the passengers face no more than 18 degrees away from the flight direction on takeoff and landing (or have head restraints), the aircraft was designed to take off without sweep, which required folding wing tips. The landing gear track was set at 60 feet.

The FAA requirement on passenger orientation is, of course, met by, and may agree with, the upper deck seat angles on Boeing 747 aircraft. One may argue that if passengers were seated facing rearward in a seat with side head cushions, a much larger angle might be permissible.

Passenger entry and emergency egress, engine and landing gear integration, as well as airport compatibility and terminal utilization were studied [428]. Four engines are placed under the passenger compartment. This configuration, because it is designed to take off without sweep, could not become the New Large Aircraft. Nevertheless it could accommodate 440 to 460 passengers and demonstrated an OFW could be designed with existing regulatory constraints.

20.4.3 NASA Studies

Waters, et al. studied the design of OFWs. They considered their aerodynamics, structures, and layout [429]. This first comprehensive study highlights several critical issues for OFWs. Principal among them were the design of the structure to carry the pressurization load, and the design of the landing gear to meet FAA taxi bump requirements.

Galloway et al. assessed the economics of 200, 400 and 500 passenger subsonic transports (M = 0.85), 300 passenger OWAs operating at Mach numbers 1.6 and 2.0, a 400 passenger OWA operating at Mach 2, and 291, 440, and 544 passenger OFWs operating at Mach 1.6 [430]. They assumed a five year development period, with 500 aircraft produced in the 15-year delivery schedule.

Aircraft prices varied from $114M to $158M for the subsonic transports, from $172M to $239M for the OWAs, and from $212M to $260M for the OFWs. For a manufacturer and the operating airlines to achieve a 12% return on their investments required 11 to 10.1 cents per Revenue Passenger Mile (RPM) for the subsonic transports, decreasing with increasing size, 12.4 to 12 cents per RPM for the OWA and 14.2 to 11.1 cents per RPM for the OFWs, again decreasing with increasing size. This trend toward economic equality between OFWs and advanced subsonic transports is depicted in Figure 144. The fares on a large OFW should compete well with the fares of its subsonic counterparts.

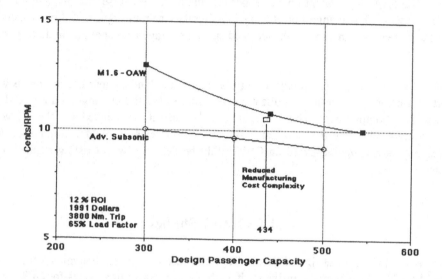

Figure 144 **NASA Ames economic assessment of the revenue required per RPM to provide a 12% return on investment (ROI) for advanced subsonic transports, and for a M = 1.6 OFW (OAW) aircraft as a function of size (from Ref. [430]).**

Computational Fluid Dynamics tools were used extensively to optimize a realistic Mach 1.6 OFW design with the wing swept to 68 degrees [431]. Related numerical studies were also conducted by Cheung [432]. This optimized design, and that by McDonnell Douglas reported in the following section, were tested in the NASA Ames 9- by 7-foot Supersonic Wind Tunnel at Mach number 1.6 [433]. The NASA design was tested at Mach numbers between 1.56 - 1.80 with unit Reynolds numbers of 1.0 to 4.5 million per foot. The angle of attack was varied from 0 to 6 degrees at a single sweep angle of 68 degrees. The 1.8% scale model included four nacelles and two vertical fins, one on the top and the other on the bottom of the trailing tip. The results of these studies are not yet published. Preliminary results indicate that the experiments validate the numerical studies which resulted in a design that, while not optimum, was realistic in layout. The experimental wing alone L/D, corrected to flight conditions at 52,000 feet, gives an estimated L/D of 10.5 [private communication, R. Kennelly].

20.4.4 McDonnell Douglas Studies

The studies by McDonnell Douglas Aerospace West have already been used to provide guidance on how many passengers a large OFW might carry, how much it might weigh, and at what speeds and altitudes it might fly. They first considered a Mach 1.6 wing swept to 68 degrees [418]. This is the design tested by NASA. They subsequently studied OFW designs for Mach numbers of 0.85, 0.95, and 1.3 over a large range of sizes. These designs were compared to point designs for a conventional subsonic transport and a blended wing body at $M = 0.85$. At this Mach number the blended wing body shows greater promise.

A subsequent study concluded that a Mach 1.3, 750-800 passenger OFW, with a 5200 nautical mile range, would require an unswept aspect ratio of about 10 [419]. With a passenger cabin height of 82 inches and a nominal airfoil thickness of 17%, the chord becomes about 55 feet and the span about 455 feet. This results in a wing area of 20,788 square feet, a volume of 109,660 cubic feet, and an aircraft takeoff gross weight of 1.6 million pounds, with 0.75 million of this weight in fuel. A sweep angle of 62.5 degrees was used; with $M = 1.3$ this provides a nominal normal Mach number of 0.6 and an estimated L/D of 10.75. This aircraft is depicted in Figure 145.

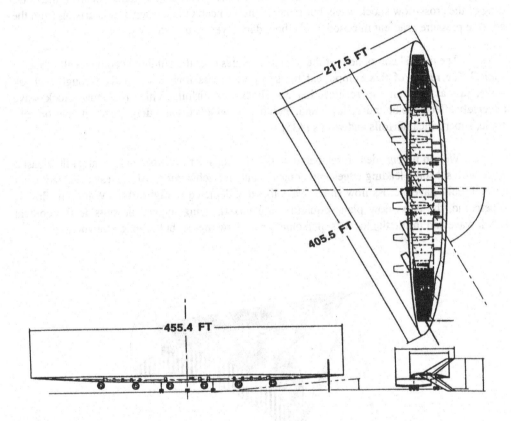

Figure 145   **McDonnell Douglas $M = 1.3$ OFW capable of carrying 800 passengers 5146 nautical miles (Ref. [419]).**

20.5 Nonlinear Theory

Following simple sweep theory, we must examine the section design of an OFW for the normal component of velocity. At cruise conditions, the flow over an OFW is that behind the nearly conical shock wave emanating from the leading tip. The wing is swept so that the component of this flow normal to the wing's leading edge will be sufficiently subsonic that a thick, shock-free air-foil may be found. We have assumed this sweep to be 60 degrees, and a freestream Mach number of $\sqrt{2}$, giving a normal Mach number of 0.707, and a tangential component of 1.23.

Boerstoel provides guidance on how thick a non-lifting airfoil might be if designed to be shock free [434]. His results, and those of others, suggest it should be possible to design an 18% thick shock-free airfoil for a normal Mach number, M_n, of 0.76. This suggests to us that a 17% thick airfoil with a c_l of 0.6, corresponding to a wing C_L of 0.15, might be shock free for a Mach number of 0.7. The importance of a nearly shock-free design stems not from the wave drag of the cross-flow shock wave, but rather from the need to avoid separation arising from the adverse pressure gradient imposed on the boundary layer by this shock wave.

The normal component of the flow accelerates over the wing to become locally "super-sonic." The return of this component to "subsonic" cross flow is normally through a shock wave, just as it is on supercritical but not shock-free airfoils. This cross-flow shock wave adversely affects the boundary layer and, thereby, the wing's lift and drag, just as it does on sub-sonic, supercritical airfoils and wings [435].

We can fix our ideas for supersonic flow by considering supersonic conical flow past a wing with subsonic leading edges and conical camber. Such a wing will, unless designed using special tools, have a cross-flow shock wave like that depicted in Figure 146. While this flow is supersonic, the cross-flow plane equations are mixed, being hyperbolic outside the conical shock wave and inside the local "supersonic" cross-flow region, but elliptic elsewhere.

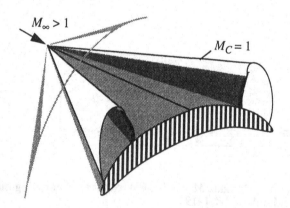

Figure 146   Embedded shock wave in a conical cross-flow.

The fictitious gas method for the design of supercritical airfoils, as discussed in Chap-ter 7, applies to conical supersonic flows as well. This extension to supersonic flows was first demonstrated by Sritharan [436]. Recently, we have suggested that this method may apply to fully three-dimensional supersonic flows [437].

The axial component of the flow over the oblique wing causes a disturbance to the

Mach number distribution of the normal component along the span. We may approximate this axial flow by that over a slender body of revolution whose cross-sectional area equals that of the wing, by using the linear theory. The resulting variation of the normal Mach number in the span direction is shown in Figure 147. We see that there is essentially a linear variation about the mid-chord value ($y = 0$) in the normal component of the Mach number, requiring different air-foil designs, or at least thickness, along the wing. We recognize, then, that the upper surface curvature and thickness of wing sections should be decreasing toward the wing trailing tip, in order to avoid creating a cross-flow shock or increasing a shock's strength if one has already formed. Our objective is to find out how well we might do in designing an OFW using supercritical air-foils that we develop and then appropriately blend to form the wing.

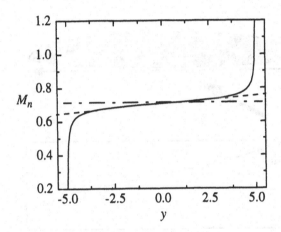

Figure 147 **Estimated variation in the normal Mach number component along the wing span; $M_n = 0.707$ at $y = 0$. Axial flow at $M_t = 1.23$, direction from left to right.**

As noted earlier, the OFW with the minimum (induced and wave) drag due to lift has an elliptic load. The OFW area distribution that minimizes the wave drag due to volume, or due to thickness, is the Sears-Haack body for volume, and the Sears body for thickness. The former corresponds to a parabolic thickness distribution; the latter to a wing that has the same center section, but thinner outboard sections. In the studies reported here we have minimized the drag due to lift, and varied the thickness somewhat about a constant thickness to chord ratio. This corresponds to a nominal elliptic thickness distribution and results in more volume, and more wave drag, than necessary. Thus these results should be considered very conservative, with per-haps twice the wave drag needed for the volume required for the passengers and fuel.

20.6 Results

At first we perform a preliminary airfoil design using the fictitious gas method. A 17.4% thick baseline airfoil is generated using the geometry tools described in Chapter 9 for a flow of $M = 0.707$ with $C_l = 0.6$. For choosing the fictitious equations used in the Euler solver, we prescribe a new energy equation to change the equations inside a local supersonic region so that they remain elliptic there [413]. This results in a shock-free flow with a smooth sonic line, but the wrong gas law, inside the supersonic region. As described in Chapter 7, the correct mixed type structure of the transonic flow is recovered in the next step: supersonic flow recalculation by means of the method of characteristics, using the just calculated data on the sonic line for the initial values. This recomputation of the flow with the correct equations of state has a lower density in the supersonic flow and provides a modified, and thinner, airfoil design. The result is the slightly flattened section shown in Figure 6 of Chapter 7 with a thickness of 17%.

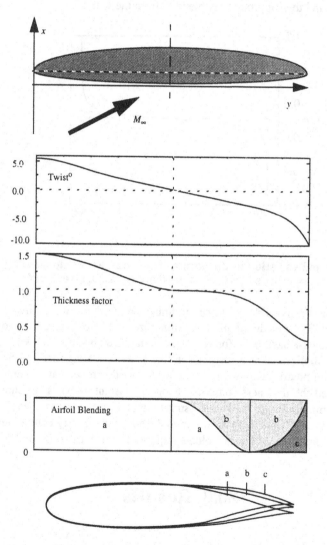

Figure 148   Wing geometry parameters: planform, twist distribution and thickness factor along span; support airfoils and their blending weight.

We use this shock-free, redesigned airfoil as the center section for an OFW. We take the planform to be comprised of two ellipses, as shown in Figure 148, with an overall 10:1 axis ratio providing an unswept aspect ratio of 12.7. Other airfoil sections are developed for slightly higher and lower Mach numbers based on the variation of the normal Mach number down the wing as determined previously using linear theory. We then use the geometry generator of Chapter 9 to determine a blending of candidate wing sections. Numerical computations were performed using NASA Langley's CFL3D [438], [439]. To achieve the elliptic load distribution, twist is varied along the wing span. The twist variation is linear near the center, strongly decreases at the trailing tip, and slightly increases at the leading tip. But different Mach numbers and sweep angles require differing twists. An elliptic loading is therefore best realized by bending the wing up at the tips. For simplicity we have used wing twist in our studies.

In Figure 148 we depict the blending of supercritical sections and the variation of twist used to achieve a nearly elliptic loading at the shock-free design point. The twist was varied from -10 to +5.5 degrees; a wing section thickness factor varied from 0.3 to 1.5 from the trailing to the leading tip. Between the center section and trailing tip, the three support airfoils shown were blended to comprise the wing. This provided an inviscid L/D of 14.36 at a C_L of 0.15. We explored the variation in L/D with angle of attack, as shown in Figure 149. We find the maximum L/D is 14.37 at a C_L of 0.146. The inviscid pressure distributions at five span stations, the spanwise load distribution, and isotachs on three grid surfaces are depicted in Figure 84. These results correct those given earlier [437].

Using the skin friction estimate provided earlier gives a viscous L/D of 10.19, which is consistent with the result given in Chapter 19. Higher values are surely possible as the designed wing's volume, 152,189 cubic feet, is considerably larger than that required.

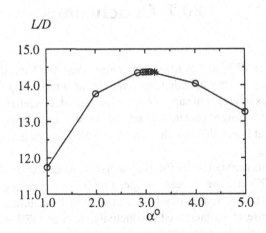

Figure 149   **Lift-to-drag ratio as a function of angle of attack, * denotes the shock-free airfoil design point.**

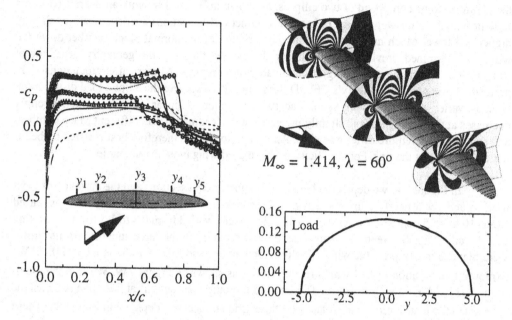

Figure 150   Pressure distributions at five span stations (y_1-dashed, y_2-dotted, y_3-solid, y_4-triangle, y_5-circle), and isotachs on three grid surfaces for OFW with elliptic load distribution obtained with varying wing sections and nonlinear twist distribution.

20.7 Conclusions

The advantages of the OFW are its low aerodynamic drag at all speeds, and its low structural weight. The variable geometry afforded by changing the wing sweep from, say, 45 degrees at take-off, to 60 degrees or higher in supersonic cruise, provides excellent subsonic, transonic and supersonic performance, low airport noise, and less concern about emissions in supersonic flight because it would fly at lower altitudes than an SCT (HSCT) in supersonic flight.

Because of its aerodynamic efficiency at transonic and supersonic speeds, it also offers the prospects of a 50% or more increase in speed on overwater routes, and as much as a 20% increase on over land routes. And it may do this at no more than current subsonic transport total operating costs because of its increased productivity. A large OFW could ultimately capture 25% of the revenue passenger miles [440].

The OFW's disadvantages are the multiple new technologies that would be introduced, its match to existing certification requirements and runway widths, the non-ideal shape for the structure required to contain cabin pressurization, the need for active control, the limited speed of Mach 1.6 or less, and the psychological impact of an unsymmetrical configuration.

Much remains to be determined about the OFW's aerodynamics, structures, and control. Leading edge computational technologies should make it possible to address the aero-servo-elasticity of an OFW. If promising results are obtained to the technical challenges of the OFW, then an experimental aircraft program is warranted to verify these findings and explore related issues. The world-wide excess military aircraft production capability could make an experimental aircraft program less expensive than it might otherwise be.

Adam Brown has pointed out that, "... as the size of an aircraft is increased, economies of scale can be obtained. But at some point, the engineer's dreaded 'square/cube' law takes over and increasing size actually results in *worse* structural efficiency. The Very Large Aircraft appears to be close to this cross-over point" [441]. This suggests, then, that we must side-step this dreaded law by considering new geometries. Among them, the OFW is the aerodynamic and structural optimum. The larger an OFW is, the higher its aspect ratio may be, increasing further its aerodynamic performance. Thus the OFW appears to be an ideal candidate for a new large transport.

It is unlikely that such a radical change in aircraft design would first occur in a large commercial aircraft. Thus an OFW is more likely to be first introduced as a military cargo or tanker aircraft, or as a smaller supersonic transport. Because of its high efficiency and productivity, and the requirement for active control, a commercial OFW might first enter service as a cargo aircraft in order to demonstrate its safety for commercial passenger service.

Acknowledgments:

This work was partially funded by the German Alexander-Von-Humboldt Foundation through a 1991 Max Planck Research Award, and by a grant from the late W. Edwards Deming. The author thanks the DLR Göttingen for their kind hospitality, April through June, 1995.

20.8 References

[408] **Von Kármán, Th., Burgers, J. M.**
        Aerodynamic Theory, W. F. Durand ed., Vol. 2, Springer, pp. 172-175, 1934.

[409] **Sears, W. R.**
        On Projectiles of Minimum Wave Drag, *Quart. Appl. Math.*, Vol. 4, No. 4, pp. 361-366, 1947.

[410] **Haack, W.**
        Geschossformen kleinsten Wellenwiderstandes, *Lilienthal-Gesellschaft für Luftfahrt*, Bericht 139, pp. 14-28, 1941.

[411] **Jones, R. T.**
       Theoretical Determination of the Minimum Drag of Airfoils at Supersonic Speeds, *J. Aero. Sci.*, Vol. 19, No. 12, pp. 813-822, 1952.

[412] **Smith, J. H. B.**
       Lift/Drag Ratios of Optimized Slewed Elliptic Wings at Supersonic Speeds, *Aeronautical Quart.*, Vol. 12, pp. 201-218, 1961.

[413] **Li, P., Seebass, R. Sobieczky, H.**
       Oblique Flying Wing Aerodynamics, *AIAA First Theoretical Fluid Mechanics Conference*, AIAA Paper 96-2120, 1996.

[414] **Kogan, M. N.**
       On Bodies of Minimum Drag in a Supersonic Gas Flow, *Prikl. Mat. Mekh.*, Vol. 21, No. 2, pp. 207-212, 1957.

[415] **Jones, R. T.**
       The Minimum Drag of Thin Wings at Supersonic Speeds According to Kogan's Theory, *Theoretical and Computational Fluid Dynamics*, Vol. 1, pp. 97-103, 1989.

[416] **Jones, R. T.**
       The Minimum Drag of Thin Wings in Frictionless Flow, *J. Aero. Sci.*, Vol. 18, No. 2, pp. 75-81, 1951.

[417] **Jones, R. T.**
       The Flying Wing Supersonic Transport, *Aero. J.*, Vol. 95, No. 943, pp. 103-106, March 1991.

[418] **Agrawal, S., Liebeck, R. H., Page, M. A., Rodriguez, D. L.**
       Oblique All-Wing Configuration: Aerodynamics, Stability and Control, McDonnell Douglas Corporation Final Report, NAS1-19345, 1993.

[419] **Rawdon, B. K., Scott, P.W., Liebeck, R. H., Page, M.A., Bird, R. S., Wechsler, J.**
       Oblique All-Wing SST Concept, McDonnell Douglas Contractor Report, NAS1-19345, 1994.

[420] **Sommer, S. C., Short, B. J.**
       Free-Flight Measurements of Turbulent-Boundary Layer Skin Friction in the Presence of Severe Aerodynamic Heating at Mach Numbers from 2.8 to 7.0, NACA TN 3391, 1955.

[421] **Peterson, J., B., Jr.**
       A Comparison of Experimental and Theoretical Results for the Compressible Turbulent-Boundary-Layer Skin Friction with Zero Pressure Gradient, NASA TN D 1795, 1963.

[422] **Galloway, T. L., Phillips, J. A., Kennelly, R. A., Jr., Waters, M. H.**
       Large Capacity Oblique All-Wing Transport Aircraft, *Transportation 2000: Technologies Needed for Engineering Design,* pp. 461-490, 1996.

[423] **Van der Velden, A. J. M.**
       Aerodynamic Design of a Mach 2 Oblique Flying Wing Supersonic Transport, NASA Contractor Report 177529, 1989.

[424] **Van der Velden, A. J. M.**
Aerodynamic Design of the Oblique Flying Wing Supersonic Transport, NASA Contractor Report 177552, 1990.

[425] **Van der Velden, A. J. M.**
Aerodynamic Design and Synthesis of the Oblique Flying Wing Supersonic Transport, Ph.D. Dissertation and Stanford University Report SUDDAR 621, Stanford University, 1992.

[426] **Morris, S. J.**
Integrated Aerodynamic and Control System Design of Oblique Wing Aircraft, Ph.D. Dissertation and Stanford University Report SUDDAR 620, Stanford University, 1990.

[427] **Morris, S. J., Tigner, B.**
Flight Tests of an Oblique Flying Wing Small-Scale Demonstrator, AIAA Guidance, Navigation and Control Conference, Baltimore, MD, August 7-9, 1995, AIAA Paper 95-3327.

[428] **Boeing Commercial Airplanes**
Cooperative Program to Develop an Oblique All-Wing Supersonic Passenger Transport, Final Report, NAS1-19345, 1993.

[429] **Waters, M. H., Ardema, M. D., Roberts, C., Kroo, I.**
Structural and Aerodynamic Considerations for an Oblique All-Wing Aircraft, AIAA Aircraft Design Meeting, August 24-26, 1992, AIAA Paper 92-4220.

[430] **Galloway, T., Gelhausen, P., Moore, M., Waters, M.**
Oblique Wing Supersonic Transport, AIAA Aircraft Design Meeting, August 24-26, 1992, AIAA Paper 92-4230.

[431] **Saunders, D. A., Kennelly, R. A., Cheung, S. H., Lee, C. A.**
Oblique Wing Design Experience II (in preparation).

[432] **Cheung, S.**
Viscous CFD Analysis and Optimization of an Oblique All-Wing Transport, NASA CDCR-20005, 1994.

[433] **Kennelly, R. A., Jr., Bell, J. H., Buning, P. G., Carmichael, R. L., Lee, C. A., McLachan, B. G., Saunders, D. A., Schreiner, J. A., Smith, S. C., Strong, J. M.**
Integrated Test and Analysis of a 'Realistic' Oblique All-Wing Supersonic Transport Configuration (in preparation).

[434] **Boerstoel, J. W.**
Review of the Application of Hodograph Theory to Transonic Airfoil Design, and Theoretical and Experimental Analysis of Shock-Free Airfoils, *IUTAM Symposium Transsonicum II*, K. Oswatitsch ed., pp. 109-133, 1976.

[435] **Sobieczky, H., Seebass, A. R.**
Supercritical Airfoil and Wing Design, *Annual Reviews of Fluid Mechanics*, Vol. 16, pp. 337-363, 1984.

[436] **Sritharan, S. S.**
Delta Wings with Shock-Free Cross Flow, *Quart. Appl. Math.*, Vol. 43, No. 3, pp. 275-286, 1985.

[437] **Li, P., Sobieczky, H., Seebass, R.**
A Design Method for Supersonic Transport Wings, AIAA Paper 95-1819, *Proceedings of the 13th AIAA Applied Aerodynamics Conference*, pp. 474-483, 1995.

[438] **Thomas, J. L., Taylor, S. L., Anderson, W. K.**
Navier-Stokes Computations of Vortical Flows over Low Aspect Ratio Wings, *AIAA J.*, Vol. 28, No. 2, pp. 205-212, 1990.

[439] **Anderson, W. K., Thomas, J. L.**
Multigrid Acceleration of the Flux Split Euler Equations, AIAA Paper 86-0274, 1986.

[440] **Li, P., Seebass, R., Sobieczky, H.**
The Oblique Flying Wing as the New Large Aircraft, *20th International Council of the Aeronautical Sciences Congress*, 96.4.4.2, 1996.

[441] **Brown, A.**
Airbus Industrie's Aircraft Development Plan and Challenges, 4th H. K. Millicer Lecture, RMIT, September 1994.